38011
THE HIGHLAND COUNCIL
KT-382-342

Sir John Lister-Kaye is one of Scotland's best-known naturalists and conservationists. He has lectured on wildlife and the environment on three continents and served prominently in the RSPB, the Nature Conservancy Council, Scottish Natural Heritage and the Scottish Wildlife Trust. His Aigas Field Centre, founded in 1977, has won international recognition for its environmental education programmes; it continues to welcome study groups from all over the world. In 2003 Sir John was awarded an OBE for services to the Scottish environment. He is a *Times* columnist and the author of six previous books, including the highly acclaimed *Song of the Rolling Earth*, published by Little, Brown.

You can visit the Aigas Field Centre at: www.aigas.co.uk

Praise for *Nature's Child*

'. . . a masterpiece, beautifully written by one of Scotland's best-known conservationists' Iain Thornber, *Scots Magazine*

'A beautifully written book that is both informative and uplifting, that reminds the reader how scary and fascinating the world looked when we were kids' Ian Valentine, *Country Life*

'Superb, a really important book everyone who cares about nature and children should certainly read. It deserves to stand alongside Edmund Gosse's *Father & Son* as a tender, lyrical, erudite story of a child's education through nature' Roger Deakin

'This labour of love deserves to be seen as a minor classic of the natural-history genre' Alan Hendry, *John O'Groat Journal*

'An entrancing utterance of controlled rapture . . . extraordinarily memorable' Magnus Magnusson

Also by John Lister-Kaye

SONG OF THE ROLLING EARTH

ONE FOR SORROW

ILL FARES THE LAND

SEAL CULL: THE GREAT SEAL CONTROVERSY

THE SEEING EYE

THE WHITE ISLAND

Nature's Child

ENCOUNTERS WITH WONDERS
OF THE NATURAL WORLD

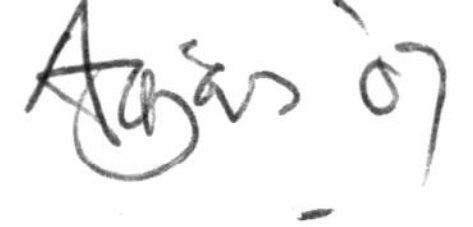

John Lister-Kaye

An *Abacus* Book

First published in Great Britain in 2004 by Little, Brown
This edition published by Abacus in 2005

A CIP catalogue record for this book
is available from the British Library.

ISBN 0 349 11760 8

Typeset in Century Old Style by M Rules
Printed and bound in Great Britain by
Clays Ltd, St Ives plc

Abacus
An imprint of
Time Warner Book Group UK
Brettenham House
Lancaster Place
London WC2E 7EN

www.twbg.co.uk

Illustrations by
Derek Robertson

We shall not cease from exploration
And the end of all our exploring
Will be to arrive where we have started
And know the place for the first time.

Little Gidding, T. S. ELIOT

For Hermione
and for nature's children the world over

CONTENTS

ILLUSTRATIONS

by DEREK ROBERTSON

ACKNOWLEDGEMENTS

My special thanks go to Derek Robertson for his sensitive illustrations; it has been a delight to work on both *Song of the Rolling Earth* and now this book with an artist (painting is his real love) who is also a most accomplished naturalist.

To our Senior Education Officer Duncan Macdonald, our Field Officer Sarah Kay, our Environmental Education Officer Brona Doyle, and the many other rangers and staff at Aigas Field Centre who, over the past ten years, have helped Hermione get to grips with the world of nature that surrounds us and fills our lives here, I extend my deep gratitude. The importance of their influence and enthusiasm cannot be understated. Susan Luurtsema was our Programme Manager during the writing of this book and she also gave me valuable support and assistance with the planning of some of our expeditions and the proofreading of texts. Our new manager, Yvonne Brown, together with Sheila Kerr, have both been a huge help in organising my life so that I can find the time to write. As always, my sincere thanks to my editor Catherine Hill and her colleague Kirsteen Brace at Time Warner Books, and to my literary agent, constant advisor and friend, Catherine Clarke.

Thanks also go to Dr Duncan Halley of Nina Niku in Norway, to Dr Göran Hartman of the University of Uppsala in Sweden, and to Paul Ramsay, Karen Barclay and Amelia Lister-Kaye for all the help I received in Norway while researching and learning about beavers; to Bobby Archibald, skipper of the good ship *Pippin*; to Tom Jamieson, our boatman to Mousa; to Ewan Masson, Nick Langton, Tim Rundle and Mick Reilly for their endless patience during our time in Botswana, South Africa and Swaziland; and to Jens Abild and Lisa Ström for their friendship and invaluable assistance in Svalbard. I extend my gratitude to the memory of the late Sir Laurens van der

Post, who first took me to the Kalahari and whose own books and friendship inspired me to return there with my family. My good friends Roy and Marina Dennis had more to do with the jackdaws Dawn and Dusk than I have revealed in the text, and their constant encouragement and support in many aspects of our work has meant a great deal to me.

Above all, without the blessing of my wife Lucy, often against all her maternal instincts, I would never have been able to share Hermione's world, nor she mine, in such wonderfully wild and sometimes quite hazardous places. Lucy's deep sigh, and her smiling resignation that being married to a naturalist is perhaps marginally better than being married to a racing driver, have become the hallmark of our many farewells before she has watched us drive away on yet another adventure.

John Lister-Kaye
House of Aigas, 2003

PROLOGUE

Beauty and the Buff-tip Moth

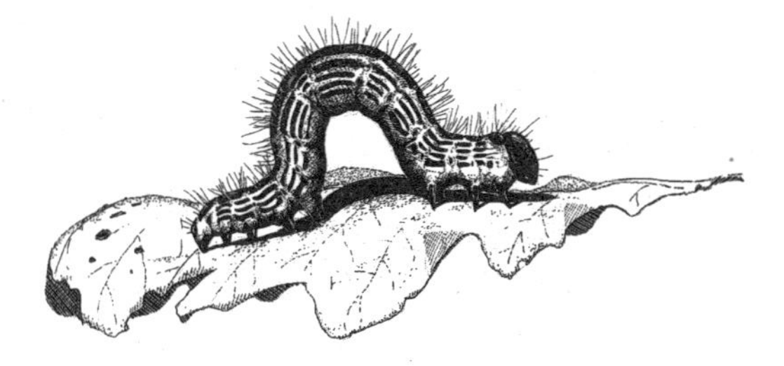

I, too, have in Arcadia been.

THEOCRITUS, THIRD CENTURY BC

It's now hard to remember when first I noticed that I had a co-conspirator, though it has only been a handful of years. I can readily bring to mind beach picnics when I sat in wonder at my youngest daughter so utterly absorbed by picking through pebbles and shells, or prodding sand-hoppers with a straw. But I could never have predicted back then that she would become my constant companion – that I would come to take her interest and her companionship for granted, to accept it as the norm. It happened slowly.

I remember Hermione standing beside a small tree, wearing pale pink shorts and a blue T-shirt. She must have been nearly seven years old. It was August, but to her it was simply summer. Her mind was fixed upon a discovery that had bent her whole attention to this one small tree. I watched her move closer, tiptoeing in as if she didn't want to frighten something, pressing her face in among the lower leaves and fixing it there for a full minute. Then she ran to the house, arms flying, urgency stretching her paces. A few moments later she hurried back to the tree carrying a plastic sandwich box.

She had found some caterpillars on a young red oak sapling in the garden. She had spotted the wispy, fleshless ribs of a single leaf. At six years old she knew the signs: that stripped leaves meant one thing – caterpillars. It was as though that day Robert Louis Stevenson's blind Old Pew had come tapping and limping in to tip her 'the black spot', inextricably embroiling her in an adventure not of her own devising, but as part of the plot, in it whether she liked it or not – no going back.

Even by the age of five Hermione was a child of the woods and the meadows that surround her home. She was often to be found in ditches, jam jar in hand; or capturing the squirming secrets beneath a log, but that day it was the specimen red oak sapling that grabbed her attention: '*Quercus rubra*', said the tag she hadn't read, 'North America'. Carefully she picked the caterpillars from the leaves and placed them in her lunch box. I didn't know it at the time, but this was the beginning of an extended exploration, which as the years unfolded was to become almost as gripping as *Treasure Island*; an adventure that she and I were to pursue relentlessly for the next seven years – those fat, Arcadian, literally wonder-full years of childhood.

For me, Hermione's discovery of those caterpillars of the buff-tip moth, *Phalera bucephala*, looking more like a lichen-covered twig than a moth, were the first inkling of the realisation that I had a fellow naturalist in the family. I watched her collect every caterpillar she could – over seventy – and secure them in her sandwich box. They were very beautiful and they had caught her eye. When they hatched as tiny hieroglyphs on the underside of one leaf, the beauty was already there – it came built in, part of the package. They arrived with intricate rings of fiery colour, their whole length squared off into curved boxes of shiny black with bright yellow margins.

Yellow and black: those two colours wave an ancient flag – one of nature's favourite ploys – a well-practised duet, endlessly repeated. They are a combination that shouts out loud. Wasps and hornets use it all the time, so do snakes and poisonous tree frogs, and a thousand other things that are either very nasty or want you to think they are. Biologists call it 'aposematic' – the warning of colours. These caterpillars also come with a large black head, businesslike and glossy, with pale hairs. In fact the whole length of each caterpillar bristles with these white hairs, as sinister as a nettle. And not content with its yellow stripes and bristling black boxes, it has a nasty smell, even before you get to the foul taste. The buff-tip moth wasn't taking any chances. Birds given to dining on caterpillars think twice before meddling with these handsome, unpalatable bundles.

But Hermione didn't plan to eat them and she knew nothing of aposematic coloration. It was the caterpillars' inadvertent beauty that had caught her eye and she was doing the very thing nature was supposed to prevent, picking them off the oak and putting them in her plastic box – a lunch box loaded with bad taste.

Buff-tip-moth caterpillars can – and do – defoliate whole young trees. They can munch the flesh from every last leaf, leaving behind them bare skeletons of whole saplings. They shuffle their intricate colours from branch to branch, not giving a damn. Why would they? They have been brought up to know that they look frightening and taste awful; no one told them they were beautiful too. Gardeners haven't been around long enough to be included in the formula the buff-tip used for its grand plan, neither have little girls with a passion for nature and an eye for beauty. On that particular tree the caterpillars were doomed, an error of judgement by the buff-tip moth. It might have been such an unhappy story, but Hermione removed them all and lovingly kept them in a large cardboard box. For two weeks her caterpillars ate their way through barrow-loads of leaves from the only other, luckily much larger, red oak for miles.

Do I tell her all this, I wondered? Do I draw her deeper into the subtle and complex machinations of science-based natural history as we now pursue it? When do you teach and when do you let life be its own tutor? Something held me back: wouldn't aposematic coloration and natural selection be wasted on a six-year-old? Might it spoil her fun, burst some special bubble of childhood fascination I had long forgotten existed? Do I wait for the questions – allow her the space to make her own discoveries?

Slowly, as I watched her, my own rural childhood came spilling back: the thrill, the absorbed, locked-on, private excitement of finding things out for myself, a thrill I had forgotten and cast aside, a sensation I had allowed to be crowded out by information and the rampant imperatives of modern adulthood. If nature was to claim this child as it had me forty-five years before her, then that was her good fortune and mine, for as long as it would last. My role was not to manipulate events or apply pressure to match some preconceived parental zeal, but just to be there and to run with it as best I could.

That afternoon I glimpsed in this late child a rare opportunity: not only a second go at parenthood within the exciting possibilities of my own subject, but also, if she really was to become even a short-term amateur naturalist, as many children are, the chance to gain a companion with whom I could revisit my own country childhood. With the benefit of age and experience, I could meld the parent and the child within. As I stood silently watching her tend her caterpillars I made a choice I have never regretted – I decided to keep my peace. Aposematic coloration? Natural selection? There would be time enough for all that.

I had been brought up in rural England before the proliferation of the motor car turned country villages into urban dormitories; in a now vanished world where a young child could walk or bicycle or ride a pony alone through country lanes without fear of lorries or louts; a gentler, now almost unimaginable world of horses and carts, cows hand-milked into a bucket, corn stooks and fallow fields. My childhood was an Arcadian paradox. Because my mother was severely constrained by chronic ill health, I was left to my own devices. The outdoors became to me a great open book of excitement and discovery, which fed an early, almost instinctive passion for all wild things. It seems likely, now looking back across half a lifetime of living in the Highlands of Scotland, that my decision as a young man to move to the far north (to a region which was in the 1960s still referred to as North Britain), to a wild and remote land of mountains, lochs and forests, was perhaps itself an unconscious escape back towards that lost idyll of happy exploration.

It is certainly the case that I always wanted my children to enjoy the freedom implicit in an outdoor country upbringing. For most of my adult life, with my family and a dedicated staff, I have run a field studies centre at a place called Aigas, near Beauly, on the edge of the Moray Firth in the Highlands. Here I brought up my first family: my son Warwick and twins, Amelia and Melanie. In a land where the gun, the fishing rod, hiking boots and the pony were conventional adjuncts to country living, they all grew up knowing and understanding the essential revolutions of nature's wild wheel. Later joined by my second wife Lucy's almost identically aged family, James, Emma and Hamish, for a few fleeting years, the six made a formidable gang of hunter-gatherers, bringing home a smorgasbord of seasonal manna from the wild: rabbits, hares, deer, trout and salmon, grouse, pigeons, pheasants, duck, or stained shirt-fulls of mushrooms, chanterelles, bilberries, blackberries and raspberries, all breathlessly dumped, muddy, bloody, squashed and dripping on to the kitchen table, by children with glowing cheeks and garbled tales of exciting encounter and mishap.

Yet, for me, those were crowded years when work often consumed twelve or fourteen hours a day, especially in the summer months, and when the priorities of parenthood were as much to do with building up the business and paying the bills as with sharing 'quality time' with my rambunctious gang of a family. In any event, their numbers and their spread of ages had banished any thought of encouraging them to accompany me in my work as a naturalist.

Hermione was born to Lucy and me after an eleven-year gap, so that by the time she was six, Warwick, our oldest, was twenty-two and our youngest, Hamish, was seventeen. The others were at university; all six up and away, busy becoming adults, preoccupied with the inevitable picking of life's rich fruits. Lucy and I were on our own again, starting afresh, revisiting a suite of parental responsibilities we had almost forgotten and, suddenly and unexpectedly, finding ourselves with more time to devote to this latecomer of a daughter – an only child with six siblings. Had she chosen to play with dolls I have no doubt that I might still have spent happy, if perfunctory, interludes with her; but from the very beginning she was uninterested in dolls, or doll's houses, or prams, or any of the other conventional amusements for little girls, all of which were present in abundance from her three sisters. No, some maverick gene in her make-up was drawing this child to nature, to beetles, newts and tadpoles, centipedes and spiders.

One day Hermione noticed that the sound of munching caterpillars in her box had stopped. She looked closely and saw that not one was eating. They were large and striking, the biggest nearly three inches long, glowing with health, but they had stopped eating. 'Why aren't they eating?' she demanded. I told her it was time to bid them farewell. 'Why?' she asked again, looking upset.

'Because they are about to pupate, and to do that they have to go underground.'

'Oh,' she said, looking sadder still. Tenderly we carried her box of plump caterpillars to the foot of the big red oak. One by one we put them on the crinkly bark. 'They will come down when they are ready to,' I told her. 'Something inside them is changing. They aren't going to be caterpillars for much longer. They are beginning the long process of turning themselves into moths; but first they have to bury themselves underground.'

'When will that happen?' she quizzed impatiently and a little indignantly. I explained that in their own good time they would burrow their way into the earth at the base of the tree. There they would pupate and sleep out the long cold winter. 'In the spring they will hatch into fine buff-tip moths and crawl their way out into the air.'

'Oh,' she murmured wonderingly. 'So then there will be more buff-tip moths to fly round and lay their eggs on more trees.'

'Yes,' I answered, 'that is just what will happen. It's what always happens.'

'Oh good!' she beamed, as I took her small hand and led her away from the red oak tree. 'Then I can find some more next summer.' I knew then that she was hooked.

ONE

The Eyes of a Child

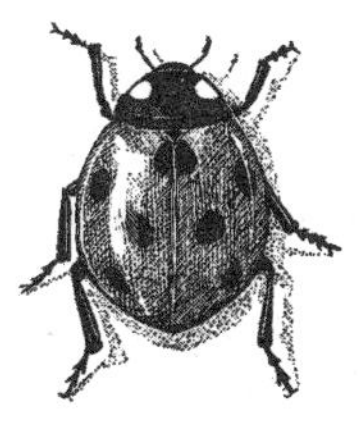

To see a world in a grain of sand,
And a heaven in a wild flower,
Hold infinity in the palm of your hand,
And eternity in an hour.

WILLIAM BLAKE, 1757–1827

For a brief moment in time Hermione and the caterpillars share a world of their own, united in nature. I stand and watch and remember, there when she needs me, unspoken support when she doesn't. But this parable is hardly an allegory for blissful existence on the planet, nor is it meant to be. It's just the happy interlocation of two human lives – a smiling acknowledgement that I can teach my daughter and by that happy process she can help me relocate the lost world of my own childhood, the memories of primal, uncluttered observation long ago, and with them, that special joy – the elation of discovery.

As I write, Hermione's twelfth year is drawing to a close. The years of innocence are waning, slipping through our fingers like fine sand, and her attention is soon to be drawn away by the hurly-burly of modern teenage life, shrill and insistent. But we have had the good fortune to live through a period of years, those from six (give me a child that can walk and talk), when a child's mind is wide open and as absorbent as a sponge, until now (and hopefully for a little longer), when other influences will inevitably barge in and take over. Six or seven blessed years of exploration and discovery, fat and full, it so happens of the natural world, because that is what surrounds her here and preoccupies her home life. It's her lot that her family home is a field studies centre among the mountains and forests of the Highlands, that her father is a naturalist who works with the kites, ospreys, eagles, otters and pine martens of a remote and beautiful glen and that my work periodically takes me to wild and exciting places around the world; hers, too, that I have often been able to take her with me – although it has sometimes seemed to me to be the other way around.

They have been years of recollection, literally the re-collection of those wonders and bright images that shaped me into who I am and how I think. Without an inkling of the effect she was having, she has forced me to remake many of the discoveries of my youth, visiting them again after an absence of more than forty-five years, with that vivid and untrammelled freshness that is the hallmark of a child's perception. Hermione has brought me to remember things I didn't know I had forgotten, caused me to smile suddenly with déjà vu, and to pick up and re-examine familiar things, a shell or a crab carapace on a bcach, a fir conc or a tadpolc, things I have scarcely bothered to think about since I was her age. I have stood beside her holding my peace as I watched stark nature ride the roller-coaster of her formative emotions – through ecstasy to agony and back again – a process that I first experienced so long ago.

For all its anthropocentrism – the shared experiences of a father and a daughter – this book is about nature and the nature's child within us both. It is about nature raw and unabridged, and, I hope, unmolested by sentiment. It is about the nature of first-hand experience heightened by the presence of a child, not so much sought out in the sense that you might search for a bird's nest, as simply found and observed in the process of life and work. It's about wild nature as it exists all round us all the time, wherever we are – in my case today, as I sit quietly scribbling beneath the big oak tree that shades a pond in the garden of our Highland home – in the frog and the moorhen, the robin and the woodpecker, and in the water boatman and the ladybird, to select only two bugs from a galaxy of invertebrates which throng around us and prompted the celebrated biologist J.B.S. Haldane to observe, 'God had an inordinate fondness for beetles.'

But for all the apparent abundance of some common species, the observations in this book also make the point that it is wild nature, which, to our everlasting loss, our modern world barely bothers to acknowledge – and this at a moment in the history of life on earth when man's actions threaten the existence of just about every life form, including our own. These common creatures also incidentally demonstrate, of course, that nature is everywhere we look – its wonders especially visible to a small child – if only we can find the time and take the trouble to seek them out. And trouble is the word. Nowadays, in most of our association with nature, trouble gets in the way. It blocks us out.

Mankind has become blind to nature, and careless of it. We shut it out of our lives, so often seeming neither to need nor understand it. Modernity deals with nature in extremes: we can't make up our mind how to handle it. We either smother it in sugary sentiment, fail to value it at all – trampling it underfoot in our blind and competitive haste – or value it falsely and excessively, ruthlessly hunting it down for our own ends.

We view the wild environment myopically, fearing worse is to come but unable to do anything about it: species extinctions, habitat loss, deforestation and desertification, over-fishing . . . the list is endless. The collateral damage of our actions swamps us with a litany of global environmental issues such as climate change, industrial and agricultural pollution, air and water quality, human overpopulation and so on, a litany that repeats like a chorus, decade upon decade. In the twentieth century we nurtured a culture of trouble with nature and then we built an industry around it.

Our response to these perceived and very real problems is an ever-increasing vicious circle of trouble, like rings expanding from a pool where humanity flails and flounders about. Finding ourselves out of our depth we have latched on to science, as if its artless disciplines were our only guiding force. In blind faith we have come to treat science as a religion, and we expect it and its rampant technology to cure us of all ills and absolve what little remains of nature within our not-so-troubled consciences. We shrug our shoulders and leave the scientists to sort things out – a cynicism, as Oscar Wilde defined it, which knows the price of everything and the value of nothing. We continue to work against nature, not with it. Because we have lost touch with it – for most of us wild nature ceased to be a personal problem long ago – we just don't seem to need it any more.

I believe this to be a bad mistake. We need nature terribly if we hope to address the threatening global problems we all face; we need it fundamentally and we need it at the core of our being. Conservation isn't just about saving tigers or pandas, or even tropical rainforests for that matter – although they are the tragic, high-profile symptoms of the malady. No, it's much more to do with facing up to the truth of our own origins – about acknowledging deeply within ourselves that we are a part of nature and that we desperately need its blessing and help. It is, after all, 'as the gentle rain from heaven . . . twice blessed': the blessings of our past and our future combined. Understanding and prioritising the needs of conservation are vital: something the industrial West has been expert at when it suits short-term profits and popular opinion, and appallingly bad at when it comes to exercising sensible precautions and tackling difficult problems.

We all acknowledge that were it not for the emergence of science and its sequitur, the great spinning flywheel of the Industrial Revolution, we would not be who we are. Many of us would have never made it to the nineteenth, far less the twenty-first century, nor would the world be the troubled place it is – at least, certainly not in global environmental terms. Our lifestyles would still be medieval and feudal; our medicine would still be magical and herbal; our transport restricted to the speed of the sailing ship and the horse; our communication largely local, considered and deliberate, instead of continental, frivolous and profligate. Most of the world's people would be primary food producers: legions of barefoot peasants garnering thin corn with scythes.

Some might be tempted to think they would prefer it that way – that the past had a golden age – but they are probably choosing to overlook the universal scourges of smallpox, TB, typhus and polio, parasites, failed harvests, malnutrition and famine, or just crushing poverty and, especially for women, the inescapable treadmill of reproductive biology the way it used to be. At best it is a romantic delusion that primitive was only wholesome and good; at worst it is probably just as well that despite the current rash of television programmes purporting to put people 'back to nature' in remote places (which paradoxically always seem to involve helicopters, speed boats and mobile phones), we can no longer properly imagine what such a life was really like. Most people fail to comprehend that primitivism and natural idealism are the luxury products of an advanced material society.

But in shedding our innocence and pursuing the faculties of beneficial science we have also inadvertently parted company with some of the delights and treasures of the natural world: those that inspired and motivated former civilisations, including our own. We have alienated ourselves and defaulted on our debt to the earth that spawned us, spurning its greatest blessings. In losing touch we have abandoned intuition and the spirit, art, magic and the sacred in nature, the very qualities from which we once assembled our values, which shaped our cultures.

Long ago the Canadian sculptor Elizabeth Muntz wrote, '. . . mankind, once part of nature, has torn itself away from the earth, spiritually bankrupting itself in the process'. She was right. In our worship of science we have gone from pantheism to pan-atheism. We have replaced wonder and marvel with measurement – hard fact, tables of rarity, so much data – allowing science to do our thinking for us, conveniently interpreting its evidence to suit our current arguments.

We devastate the world's great ecosystems without a thought for their life pyramids and food chains; their unique species and their abstract benefits, such as beauty and spiritual renewal; their many associated intangibles, such as the birdsong that colours our days, or the long-term effect our actions might have on climate or soils. The most wonderful symbols of wildness and our own origins, the tiger and the panda, the elephant, the lion and the wolf, we encircle with lines on the map, fences on the land and quotas in the mind. We too often remain blind to the quintessence of their wildness of which we were so recently a part.

After half a lifetime of working in environmental education and conservation, I find myself fervently praying that somehow we can build on the excellent work already being done by global conservation groups to spawn a generation of youth who turn away from endless consumer capitalism because they see and comprehend the plight we are all in. I pray that one day, public opinion will force politicians to reassemble their spiritual and environmental values so that they can redirect the stark truths science now so regularly delivers up for the benefit of the whole planet.

This book is no threnody, nor is it a rant – rants may diffuse passion from time to time, but they seldom win arguments. Besides, thanks to this beautiful place in which I work, the northern central Highlands of Scotland, and to Hermione and my six adult children who really do care about our planet, my mood is surprisingly optimistic. The human spirit possesses that great capacity for hope – one of the fundamental differences between our species and just about every other life form.

As I have watched Hermione and thousands of other school children passing through our field centre, children happily learning the wonders of nature, I have found that embodied in their buoyant youthfulness I can perpetually draw strength from what Kathleen Raine has called 'the bright mountain behind the mountain', and the hope that one day we shall arrive there. Far from being doomladen and maudlin, through our work I have been privileged to rediscover many of the lost delights of nature and its values on my own doorstep, even at my side. I have felt like the luckiest man alive. Not only do I live among upliftingly beautiful mountains, forests, lochs and glens, but I have also been able to spend my professional life applying observation and the now universally founded principles of ecology to what I see about me every day. And yet I have been able to do so at arm's length.

I am very grateful that life has not tipped me down the slope of formal science; I should not have been comfortable there. I find it wanting in its obsession with measurement, identification and classification, failing to deliver up the fulfilment that I, personally, have always sought and found in nature. I can do no better than the well-worn apothegm – for me, many scientists seem to be so busy counting and measuring the trees that they fail to notice the wood. But in understanding at least how elementary science works, and applying it selectively and tentatively, I have discovered that in nature and wildness the human spirit can find tranquillity and can relocate at least some of those values that have sustained us as a species for so long.

If I have a problem with science and the techno-centric modernity it has spawned, it is certainly not a sustainable argument. My lifestyle depends as much as anyone else's upon the essential benefits we all enjoy. My air-conditioned four-wheel-drive truck takes me to remote places from where I can run my affairs with a mobile or satellite phone. I depend upon email; I cannot imagine life without my laptop, which allows me to write under a tree beside a pond. My office is cluttered with cables and technical wizardry (the functions of which bewilder me), and my colleagues and I waste far more paper, energy and other natural resources than I care to admit, even though we do try to work to a responsible environmental charter. Almost all the outdoor work on our field-centre property is achieved with power tools: tractors, diggers, mowers, quad-bikes, chainsaws, electric screwdrivers, drills and the like, and there is always some new bit of kit I am told we cannot possibly manage without. The supermarket remains seductively convenient, although I know it disadvantages small rural businesses and fractures communities, and, despite dabbling with solar power, the principal heat for my home and our field centre during our long northern winter still comes from fossil fuel. The buzzword 'sustainability' echoes uneasily around here.

I suppose that my discomfort with technology is not so much with its many obvious benefits for us all, as the way science has singularly failed to jolt the world's political leaders into seeing where we are heading. All they seem to be saying is, 'Don't worry, we have a newer and smarter computer system arriving soon.' I have come to view rampant materialism and consumerism as a socio-cultural pathology, even ultimately a fatal infection – fatal, that is, for most of the wild nature we purport to love, if not ultimately for our own species. Along with religion, culture and education, nature also seems to be rendered down to the lowest common denominator: as though we want it to be something tamed, safely quantified and packaged, there for convenience and entertainment, ever more banal.

So, in exploring wild nature with Hermione, I have sought to look beyond the problems and the failures of our scientific age, to explore some of the much more ancient benefits of nature, which until so recently lay at the heart of our lives. I mean, of course, the joy, the sense of belonging and the inner peace and fulfilment nature can deliver for us all. I include the heightened sensations we now seem to rely upon television to provide for ourselves and our children: the thrill, the leaping emotions, the fear, the humility, and the sheer, gut-clenching mortality that some real encounters with wild nature can deliver, encounters which, for all their potentially life-changing consequences, I would not wish to have omitted for me or for any of my children – experiences that allow us to know we are alive.

'What's that?' my daughter demanded one day, pointing to one of several birds gorging on scarlet rowanberries in the garden. 'They're thrushes called redwings,' I told her. 'They've just flown in from Scandinavia.' I passed her my binoculars and she studied them for a moment.

'I can't see any red,' she insisted, as though I must be wrong.

'It's under the wing,' I reassured her. 'You can't see it unless they fly.' Another pause while she considered this new intelligence.

'Then they should be called red-underwings.' She announced this with more than a hint of indignation, almost as though I'd been caught out, along with whoever first named the bird, conspiring to mislead her and the rest of the world. 'And how can it be a thrush if it's a redwing?' she added, lunging home. I laughed; a child's logic is not readily appeased.

'You know very well that a rook is also a crow,' I parried, enjoying the game. At that moment a rabble of redwings and fieldfares flew in and disrupted those already in the small tree. There was a flurry of chattering cries. Their tawny underwings flashed like spilled port. A smile flickered across her face, the smile of information well logged, but *her* information, arrived at on *her* terms, stemming from *her* curiosity – my turn to smile.

One cannot unlearn knowledge, nor even set it aside, so being forced to see nature once again through the eyes of a child has been profoundly cathartic. I have used Hermione's responses to revisit and reassess my own perceptions and experiences in nature, to see them altogether more purely and directly, which has often been great fun. I have watched her find joy in nature before it was defined for her, watched her play happily in the plunge pool of unknowledge.

This morning I looked out of my study window and saw a red kite floating and tilting past on the wind. But I could not make this simple observation without remembering that this bird was poisoned and persecuted into extinction here in Scotland as recently as 1880. Nor can I deny that it is entirely thanks to practical science that we have learned how to take surplus fledglings from nests in Sweden and hack them out as wild birds in a new place like Scotland, or Windsor Great Park or the Chilterns, or for that matter almost anywhere of suitably varied habitat for this elegant, fork-tailed, largely carrion-eating hawk. The process is called 'reintroduction' by the ornithologists who have perfected the technique. It is applicable as much to the white rhinoceros, to European beavers and to native woodland as it is to the red kite and the white-tailed sea eagle, the only two bird species so far successfully reintroduced to Scotland. Reintroduction is an essential component of restoration ecology – many ecosystems can't repair fully without their essential keystone species – and one we are likely to use more and more if we are to tackle the frightening global environmental challenges we face.

Once released, the individual red kites are monitored with tiny radios super-glued to a tail feather until the battery runs out or the bird moults the feather. Earnest young researchers with radio antennae and dishes follow the kites as they feed, mate, nest and roost. Every movement is mapped and plotted on a database. Large, coloured wing-tags with visible numbers are securely attached so that yet more enthusiastic recorders with binoculars and telescopes can identify each bird for their burgeoning records. When they breed, their chicks' legs are ringed shortly after hatching and they are wing-tagged before they take flight.

We know a great deal about the red kite, but, as it banks and wheels over the sunlit fields in front of my home, I know, too, that my view – my own cerebral perception of this handsome raptor – has been systematically manipulated by such excellent, carefully researched and considered scientific facts. My emotions are held in check. It doesn't do for even a reluctant amateur scientist like me to allow excessive joy to interfere with observation. I write down as I have been trained to: '*Milvus milvus*, no. 32. Mauve wing-tag, flying south-south-west at 09.43 hrs. 6.4.2002. Aigas river fields.

Slow, circling flight. Alone. Wind light, dry. Disappeared from view at 09.57 hrs.' I have raised my faithful binoculars and joined the game – the endless accretion of information in search of knowledge.

If I chose to pick up the telephone I could know in minutes the origin, sex and age of this kite; where it has nested and with which partners, how many chicks it has successfully raised to date; carrying what prey species to the nest to feed its young – all this would be available to me, as well as how far afield it roams in the winter months, and where and when it was last recorded. I would then feed my own information into the system. I often contribute like this and enjoy doing so.

But sometimes I have allowed science to take over my perceptions and, I fear, occasionally swamp my imagination. Somehow, mystery, wonder and awe get dumped along the way. Swapping facts and observations with my scientific colleagues is fun and I have gauged my language with them accordingly, carefully avoiding such expressions as beautiful, uplifting and life-enhancing, just in case I am thought to be contributing something frothy and superficial, later to be tossed aside as anecdotal evidence. Science and scientists too often seem uncomfortable with emotions. Aldo Leopold (there was never any doubt about his emotions) observed that knowledge condemned ecologists to 'a life of wounds', but, as was his point, they are wounds for the most part self-inflicted. There is little place for wonder in science.

Hermione doesn't see life that way – yet. Nor, when I call her to the window, does she care that the red kite is cluttered with a complicated history. She eyes it through my binoculars in silence. She steadies herself on the windowsill; I can see that she is following it closely, tilting her head this way and that, imagining that she is the kite, wheeling with it, sifting through invisible strata of sun-loaded breeze. If she sees the mauve wing-tag she doesn't mention it. Nor do I. I want her to make the most of her own observations, to hear what she feels and needs to know before fogging her with technical information. 'I can see why it's called a kite,' she says.

I wanted Hermione to make her own discoveries, as I had done, even if that involved an element of hazard. I saw it as a special role I could perform, both her tutor and her guardian, watching her joy and sharing her fears, hopefully being there in time to prevent disaster. I had had to discover some things the hard way.

At about the age she is as I write, I was given a canoe. It was an ageing kayak made of canvas shrunk on to a wooden frame. To me it was treasure untold. It was liberation from adult control and from the physical constraints of the riverbank and the lake shore. It allowed me to explore places I had long wanted to: reed beds, marshes, sand bars and spits, and to pry beneath the bank where I knew that many secrets of the natural world were hidden. I was made to wear a life-jacket – reasonably enough, even though I swam well by that age – and I was restricted to inshore waters, but beyond that I was free. Little, it seemed, could go wrong. If I capsized I swam ashore or floated downstream until I could make a landfall.

A favourite launch site in those days was the estuary of the River Axe in south Devon. These muddy tidal reaches were full of curlews, herons, ducks and cormorants, with endless cuts into salt marsh where rowdy black-headed gulls nested. On my third or fourth exploration of this exciting river world I headed upstream towards the extensive reed beds I had spotted from a road bridge a mile or two inland. The river was much narrower here and soon the tall *Phragmites* reeds fringed both banks, hemming me in. It was a new world, secret and exciting, where coot and moorhen slunk furtively in and out of secret channels, and water voles plopped and dived from their burrows. I was David Livingstone discovering the upper reaches of the Congo and I confidently expected to have to dice with hippos and crocodiles at any moment.

It was not to be crocodiles I ran into, as it turned out, but swans. As I rounded a bend I came upon four mute swans gliding serenely towards me. They didn't seem alarmed, but they quickly turned and swam away from me, upstream, their heads swivelling on elegantly looping necks as they kept a wary eye on me. I gave chase. The faster I paddled, the faster they swam. I was surprised by the turn of speed they could produce when pressed. They disappeared around a bend. When I saw them next they had gained on me and were well away, a hundred yards or more, in mid-stream, plying powerfully up a long straight stretch of river not much broader than a wide two-lane road, and still hemmed in by high reeds. I continued after them, paddling surely up the middle of the river, but slower now because the going against the current was hard.

When they reached the next bend, instead of disappearing around it as I naïvely expected, they turned towards me. Too late I realised what was about to happen – was now happening before my eyes. They extended their huge wings, rose up in the water and came at me, all four in tight formation, consuming the whole river from bank to bank, black feet paddling frantically, great wings thrashing the surface into a frenzy of white water. Each bird weighed thirty pounds and had a spread wingspan of eight feet: thirty-two feet of wildly flailing wings, each one strong enough to break a man's arm or knock me unconscious, all careering directly towards me at thirty miles an hour.

Adrenalin screamed in my arteries. Panic churned in my gut and my heart pounded. There was nowhere I could go, nothing I could do. They bore down upon me in a great tumult of white whirring wings, each vibrating beat yelling 'Fair! Fair! Fair! Fair!' – a rushing spearhead of monstrous birds, determined, pretty unfairly it seemed to me in the few seconds I had left, to smash to smithereens my precious little craft of cloth and frail timbers, and likely me with it.

The jumbo jets of the bird world, mute swans can only rise very slowly; they need a long runway. They had no choice but to run me down. They too were hemmed in, unable to rise above the reeds. I had forced their hand, pressed them a bend too far, into ever-tighter water, where they were ill at ease. They felt trapped. Inadvertently I had threatened them into panic. There was nothing else for it – I did the only thing I could. I dropped my paddle and rolled the canoe, crashing out sideways into the cold current, ducking beneath the surface just as they swept over me inches above the upturned kayak. Spitting and gulping I bobbed to the surface and clutched on to the painter from my stricken boat. The birds were far away, a high, pale streak wavering through a grey sky, heading to the safety of the broad estuary. I, too, was safe and much wiser; a lesson well learned.

As a father of now adult children I knew very well that if Hermione was to indulge the wild as I, and they, had done in our time, she too would meet with danger, know fear and panic and face the awful possibility of harm. I believed then and I do so now, that there is value in these sharp moments of trauma and their chill wake of raw emotions; that the spirit of adventure is a part of life itself and that to deny it is to place an obstruction at the very core of knowledge and understanding.

For as long as Hermione would allow me to I would be her protector and her guide. The alter ego of my distant childhood was flooding back, filling in the blanks, reawakening excitements of long ago. For the most part I had been on my own. I was determined that wouldn't happen to Hermione. That she would test me I was sure; but I both wanted and needed to share the rewards and to be there to pick up the pieces when things went wrong.

TWO

Beauty and the Beholder

> The indescribable innocence and beneficence of nature – of sun and wind and rain, of summer and winter – such health, such cheer, they afford forever! Shall I not have intelligence with the earth? Am I not partly leaves and vegetable mould myself?
>
> HENRY DAVID THOREAU, 1817–62

I sit leaning against an old pine tree with my notebook and pen on my knees, my laptop beside me. The ground is still moist from six months of sodden winter, although the air is fine and the sun generous. It is spring, there's no doubt of that: the light is yellow and the birdsong vibrant, a gift of both music and dance. Even if I didn't know it was April, if I had just awoken from a deep coma to find myself bewildered and totally blind, my first unmistakable impression would be of spring. It assaults you. It rides the breeze that teases the surface of the loch and whispers in the willow shrubbery at its edge. You can taste its tangy intoxication on the air and its irrepressible bird chorus envelops you as in Caliban's haunting dream: 'sounds and sweet airs which give delight and hurt not' and 'a thousand twangling instruments that hum about mine ears . . . that when I waked after long sleep, I cried to dream again'.

Spring seems to be surging up at me from below. Green shoots are piercing the tufts of winter-bleached grass and worm casts have oozed to the surface like toothpaste, as if the soil is sick of being stuck below ground. Today, nature is shaking out her skirts; the land is awake and panting, breathless after an early morning run. It's gone to her head; she's dizzy, on a roll. In the pines and birches around me blackbirds, thrushes, great tits and coal tits, wrens, dunnocks and chaffinches are fluting, sawing and hammering out their rousing refrains of life and duty. They are grabbing the moment as if their lives depend upon it, which, come to think of it, they probably do. It's infectious. Whatever the day is doing to my blood corpuscles is happening to theirs too, automatically. For them no thought process exists; this is no chance encounter with a bright spring day, it's programmed in. For all their melodious accomplishment they are at it in earnest, responding, bursting out and dragging me with them. They are settling old scores, fixing space for themselves and their future. Hormones are teeming; breasts swelling. Forces have combined to demand – to insist – upon this spontaneous celebration of the year's turning.

Across the other side of the loch, in the marsh, mallard are chasing their own destiny through the shallows. I can hear them and see them – at least, I can hear frantic quacking and see the flurrying among the rushes. Every now and again a skirmish of wings rises a few feet into the air and drops down again, out of sight. I feel sorry for the females. It is a duck's lot to be hounded by two or three drakes at once. If I were to creep up, as one easily could at this time of the year – so intent are they upon this pressing affair – I would see the poor duck ignominiously shoved below the surface by each swain, who grips the feathers on the back of her neck in his bill and treads her back with his bright orange webs, while his frantic tail spreads to a broad white fan. In his passion he entirely smothers her. Ducks have been known to drown this way. No sooner is he spent than another ardent drake leaps on and roughly dunks her under all over again. Spring is here all right.

I have come here to work. Once my senses have settled down a bit I shall open my laptop and (a little more delicately than the mallard drakes), I shall try to feed the spring directly into its mysterious grey belly. I need to spin words like a spider spins a silken thread from its tail. Perhaps that's what I'm doing? Like the spider I am tipping my metaphorical abdomen into the breeze and letting a slender thread of words pick me up and transport me to a new level of the forest – literally carting me off, giddy and sublimely content. But there is another reason to be here: I am on duty.

Hermione has begged me to let her launch her little boat on the loch for the first time this year. It was a present a year ago, for her seventh birthday: a grey inflatable dinghy with yellow-bladed aluminium oars fixed on articulated pins so that she can't lose them. It is eight feet long and it skims over the surface like one of the water boatmen in her aquarium. She nags and I have delayed, waiting for a day like today. Now I'm happy to concede – it's hardly a chore. I love coming to the loch and I have little fear for her safety. Not only does she swim well, but in her life-jacket – the same gay orange as the mallard's feet – she and her little boat are unsinkable. Besides, in only eight bright acres of reflected sky she can never leave my vision; even her mother is content. So off she goes, tugging at the oars and bouncing through the sunlit ripples and the birdsong, neither knowing nor caring what elemental forces are gathering at the loch's soggy shore.

In *The Once and Future King*, T.H. White recommends that all children between the ages of four and eight should spend three nights a week sleeping in a stable or a dog kennel, so that their immune systems are exposed to a sufficient range of diseases to produce lifelong resistance. To many parents that would seem the irresponsible proposition of an eccentric bachelor (which he certainly was), but, of course, it also contains a germ of practicality. The over-protection of children inevitably denies experience – a truism as much of antibodies as of responses to the hazards of everyday life. But however philosophically one views children's activities, parental anxiety doesn't go away. The 'what if?' factor is often very hard to suppress. In our pursuit of wild nature there have been many occasions when I have worried that my liberal approach to Hermione's exploits – perhaps the product of my own largely unsupervised country childhood long ago – may have placed her in danger. But here, at her home loch, with sensible precautions in place, I can relax.

I settle to my work. I open the lid of my laptop and nudge it out of its slumber. Words stream in a babble of scarcely coherent impressions. Such is the wonder of this technological wizardry that I can tip them in at random and sort them out later, entirely confident they will still be there. From time to time I glance up. I see the boat zigzag about, now on one side of the loch, now on the other, now heading for the marsh so that the mallard are forced to move further away. I am watching – I *am* fulfilling my paternal duty – but not really observing. My brain logs the distant boat and its orange blip of an occupant, but nothing more. But I do know exactly what my daughter is doing: she's exploring in the way that only a child can, places she has often explored before, finding them again and again, adding jigsaw pieces to the growing picture in her head. She is unconsciously garnering images and soaking them up, storing them away – like I am doing, but without having to work at it – to sort out later. Her thoughts are as zigzagged as her progress; she is lost in that great treasure trove of mental and temporal levitation a child calls adventure.

I'm not aware of time passing. I scroll back through my pages: an hour must have passed, maybe more. The boat is still in the marsh but Hermione is some distance from it, bent almost double. She seems to be fishing for something in the shallow water. I home in with my binoculars. Cupped in her hands she carries the thing back to the boat, lolling on its air-filled bulwark she leans in. I can't see what she is doing. She is happy; that is all that matters. I go back to my work. The next time I look up I see the boat heading back towards me. Every few strokes she looks over her shoulder and adjusts her direction with a tug. I'm torn. I have enjoyed this wafting meditation in such a beautiful place; I don't want it to end. Whatever it is she has found is being rowed across the loch for me to see. I can tell how important it is by the urgency of her stroke. I press 'save' and gently close the mystery box down. She flings me a rope. What greets me instantly snaps me out of my introspection. 'Good God!' The words are out before I can stop them.

'Aren't they lovely?!' Pure delight radiates across her face. I can scarcely believe my eyes. The bottom of the boat is alive with toads.

Quite frankly, 'lovely' is going it a bit for a toad. I concede that to another toad, especially in spring, 'lovely' might be appropriate, but there it ends. The common toad, *Bufo bufo*, is the warty and, for most of the year, dry-skinned but amphibious cousin of the frog. Placing this creature on your hand and facing it, eyeball to eyeball, reveals a fascinating animal, but it is *not* lovely. Beauty is singularly absent. Unlike the buff-tip-moth caterpillar, beauty does not seem to have been a criterion in the design of this animal.

It has a broad, slightly upturned mouth with a bony rim. Pump-like, its leathery throat pulsates rhythmically. With bright golden orbs and dark horizontal pupils, its eyes appear to be struggling to burst out of the top of its head. Its expression is a bemused sneer, as though nothing really matters, least of all you. It has no neck. Its body is plump but not fat and resides within a loose, variously dirt-coloured, golden or greenish skin covered in lumps and warts that, on a human, would indicate decades of both leprosy and smallpox. Its underbelly is palely speckled and smooth. Stubby forearms are inward turned and bowed; grubby little fingers on each hand point at each other. The double-folded rear limbs are tucked beneath its body with frog-webbed toes, less extended than its fully aquatic cousin. It prefers not to jump. It plods – in fact plodding is its thing.

Every April a ritual takes place that lies at the ecological heart of spring and freshwater wetland. It is a rite of passage, a life force inextricably linked to the turning year, which happens throughout Europe – in fact wherever there are common toads. Once seen one is unlikely to forget it. It is one of nature's many excesses and has about it something chill and Stygian, spookily mysterious in its determination.

The common toad is an insect-eating amphibian of damp habitats. Along with many other species of frog and toad worldwide it belongs to the zoological order *Anura*, but then parts company with the proper frog by having its own toad family, the *Bufonidae*. British frogs live in or very close to water all the year round; toads don't. Just like frogs toads spawn in the water to produce tadpoles (literally toad-poll; toad head), which gargle to breathe and then consume their own tails. A bit later they grow legs and crawl ashore to moist places. They lurk beneath mosses and ferns, under stones and in crevices. Coldly they creep into other people's dark burrows. I once knew a huge toad that lurked for many years among bins of vintage port in a dark, damp cellar.

When ashore and growing into adults, toads disperse far and wide. For a creature of slow movement it is surprising how far they travel. Unlike the frog they only jump reluctantly. So they plod, up to a hundred yards a day, often for over two miles from where they were born – a fairly astonishing distance.

Having found their desired damp hollow, they take three years to grow and become sexually mature. Having croaked and crept into their chosen cave; having forgotten about water for three years and grown fat on flies and bugs, slugs and worms; having laid up in clammy torpor for three consecutive winters, having done all that up to two miles away from their birth water, as soon as the air temperature reaches 4°C in their third spring, they wake up, turn around and plod all the way back to their natal pond.

Early every spring – here it happens at the end of March – toads desert their drystone walls, abandon their soggy culverts, crawl out from their dim crevices and come home. No matter how far it is: no matter whether they have to cross mountain, moorland, field or stream, whether through forests and wet woods or over rock, tarmac or pavement, clambering over the corpses of their comrades squashed on roads, at whatever cost, fixed and resolute, they plod back.

Nor can you put them off. It is this saturnine determination that I find so remarkable. Higher animals, like a mouse or a roe deer, or most birds, for instance, if disturbed in the process of giving birth or nesting, take off. They desert, often abandoning their site and their young. It takes them a little time to get their act together again. Not the toad. If you pick him up, turn him around and set him down again, he immediately reverts to his previous route. If you cart him off to somewhere new and plonk him down, within minutes he orientates back and with that same stoical grimace he plods off on his slow, methodical way.

As the end of March approaches the land here is heaving with thousands of toads. They are everywhere, in a huge ring, a circular wave of warty amphibians slowly imploding to a bright, watery core. Research has revealed that the level of glycolic acid emitted by certain algae in a water body enables the toads and frogs to *smell* their way home, like a salmon smells the run of its own river where it meets the sea. Incredibly, the concentration of glycolic acid detectable by toads at a three-quarter-mile range is only one part per eighty million. It's hardly surprising that they became associated with witches and black magic. To walk around the loch at this moment in the spring, before the toads actually take to the water to breed, especially at dusk, is to see and hear them everywhere. The air is burdened with the muted, croaky exclamations of males persistently calling out for a mate. They seem to be gulping 'Help!' with the failing cry of a drowning man. You can't take a step without treading on them. It is as though, like Pharaoh, we've committed some heinous sin and Moses has visited a great plague upon us: they bear all the gravity of a portentous omen. At nightfall the croaking becomes oppressive, almost threatening, an inescapable hoarse chorus that reverberates deep within the skull. No, lovely is not the word.

But in the streaming light of a sunny afternoon this sense of menace is absent, especially to an eight-year-old country girl. The toads have taken to the water and are in the process of breeding. In a moment which should be private and uninterrupted they have fallen captive to the over-enthusiasm of a child enthralled by her own discovery. As with the buff-tip-moth caterpillars, she is not content with one or two; she wants them all. They are a challenge and a joy. She has caught them in the act of raw, amphibian sex – called *amplexus* – the smaller male riding the female, clasping her round the waist with his bowed forearms, locking into position so that he can release his cloacal sperm on to her gelatinous bootlace of black-dotted spawn. The she-toad winds her mechanical route about the shallows, weaving her egg-strings in and out among the rushes and weeds, while her male passenger emits a slow flush of sperm into the water surrounding them. He is in a protracted orgiastic trance; she is hell-bent upon duty. They have no choice; it is for this that they doggedly plodded all the way home. So there they were this morning, dotted about among the rushes in shallow water, oblivious and preoccupied with their moment of procreational bliss – and along comes Hermione.

Tenderly she gathers a pair up. The first thing they do is to urinate on her hand. 'Lovely' seems ever more remote. Unfazed, she peers into their bulging, unblinking eyes with horizontal pupils and lovingly places them into her upturned sweatshirt. She captures another pair and another. Soon her sodden shirt is bulging, so she

wades back to the boat and decants her treasure on to its rubber floor. With the bailer she swills in two inches of water. Then she returns to the marsh for more toads. For over an hour she has gathered up hundreds of coitally embracing amphibians. She has achieved a crawling, croaking, eight-foot boatload of fecundity. Three hundred toads are sloshing around her as she climbs in, easing her feet between them so that she can take up the oars. Then she sculls back across the loch to show me her squirming, spawn-strewn consommé of unexpurgated sex. She is beaming all over her face.

At moments like this I have to take a deep breath and think hard. It would both be deflating for Hermione to be told to put them back herself and risky for the toads. Somehow I have to scoop up her over-enthusiasm and redirect it. I also have to sort out my own revulsion. However much I don't want to climb into that slimy, heaving and croaking boat, I have to. Rowing back to the marsh with her will give me the time I need. I take off my shoes and socks and, still feigning enthusiasm, ease my feet into the cold, crawling, gelatinous soup. I row while Hermione sits in the stern gazing lovingly at her charges.

Nature can be absurdly bountiful. It commonly produces wild excesses to guarantee the survival and expansion of its species; especially those lower down the food chain and the complicated nutrient pyramids and support mechanisms of ecology. Peregrine falcons and golden eagles teeter somewhat precariously at the top of their Scottish upland ecosystems. In good times they do very well. Peregrines expand their range and fledge up to four chicks a year;

eagles two. But in bad times they both fail dismally. If feeding is poor they often produce no chicks at all and those they do may not reach adulthood. Nature compensates for this by allowing the adult birds to live a long time, perhaps fifty or sixty years in the case of eagles, during which time each adult should have sufficient good years in which to replace itself several times over.

In the middle of the pyramid important prey species like the mallard, for instance, are forced to be prolific. They lay a dozen eggs at least twice a year. If the eggs get raided they start all over again. The eggs and the fluffy chicks feed just about everything that needs a tasty mouthful: trout, pike, otters, pine martens, stoats, weasels, foxes, badgers, owls, buzzards, kites, herons, gulls, crows, magpies, ravens, adders and even large eels all regularly have a go. It is a fate that scarcely bears thinking about, guaranteed to reduce a child to tears in seconds.

Many times we have watched a mother duck proudly toddle her brood to the garden pond, or witnessed a just-hatched flotilla set out from the bank of the loch to cross to the other side: a dozen or more tiny balls of cheeping fluff immaculately designed to arouse the emotions. Enter the villains: a gull or a hoodie crow. In seconds the score is down to eleven and falling. The next day thc duck has only cight – then six. By now the children are at screaming pitch, unable to comprehend why this is being allowed to happen. Finally, two are left – probably the number their mother can actually protect and keep so close that they have a chance of outwitting the ravening horde of predators that throng her, day and night. Little wonder she must produce at least two broods a year and survive for several years as a breeding female to be sure of replacing herself in perpetuity.

The toad and the frog are lower down the chain still. They are restricted to wetland habitats, especially at breeding times, and are virtually defenceless. The frog has a powerful leap, immortalised by Jeremy Fisher's close call with a trout, and the toad has nasty secretions from the warts on its skin. An inexperienced dog who tries to pick up a toad in its mouth doesn't hold on to it for long. I have seen my Jack Russell terriers foam at the mouth for half an hour, tremble all over and roll pathetically in the grass trying to rid themselves of this foul, acrid toxin burning their mouth and throat. But, despite this trick, the toad is vulnerable and gets hit hard by predation in every stage of its life. Herons, initially immune from the foul taste because the toads don't get as far as their gullet, seem to delight in skewering them with their long, sharp bills. Often the marsh and shallows are littered with stabbed corpses abandoned in disgust. If they were frogs, they would be gobbled whole. It must be that the toad's frog-like shape triggers the same predatory reaction, the protection of the foul taste coming too late to save it. Otters and badgers nip them playfully, flinging them through the air; gulls pick them up by a back leg and drop them from a great height: neither have any intention of eating even one.

Tadpoles and metamorphosing toadlets in the aquatic environment are also fair game. Everything has a go. One particularly savage predator will munch his way through every last tadpole in Hermione's annual aquarium. Each spring she imports frog and toad spawn from the loch or the garden pond into her large glass tank. She spends happy hours fishing for newts and water beetles, dragonfly and caddis-fly larvae, snails and funny twitching nymphs that fall to her sweeping net. Frog spawn she deposits in great glutinous clumps. Delicately she threads the toad spawn through the weed she anchors to the bottom with stones. Water boatmen skate across the surface; palmate newts hang weightless in the yellow, side-lit water.

In no time at all she has a captive ecosystem of her very own: growing, performing, feeding and metamorphosing inches from her nose pressed up against the glass. But if by ill luck she has imported the larva of the great diving beetle, she will watch with horror and panic. This powerful larva has jaws from hell. Huge pincer-mandibles like hypodermic syringes seize and hold the feebly struggling tadpoles while they are injected with lethal digestive juices. The tadpole is then liquidised internally and sucked hollow through the same pincers. Its empty, deflated skin sinks to the bottom like a shred of rotting vegetation. The terrifying insect will seize and suck its way through the whole tank, in a week there will be no tadpoles left.

By the time we reach the marsh my bare feet are a tangle of slime. I have persuaded Hermione that the best thing to do is to put the toads overboard slowly and carefully as I row around, making sure that they are returned in the same sort of spread and density as when she found them. This is a huge success. She is able to pick up each pair in turn, smile into their horizontal pupils and give them a name. We have Jack and Jill, Fred and Wilma, Kermit and Miss Piggy, Pinkie and Perkie . . . then she begins to run out of names. I offer Hansel and Gretel, Bonnie and Clyde, and Romeo and Juliet, which go down well, but Porgy and Bess, Crosse & Blackwell, Gilbert and Sullivan, and Victoria and Albert are lost on her; she rejects them as though I don't properly understand the game.

After five minutes we still have over a hundred pairs to go. I can see this is going to take a while. Luckily the naming quickly palls and we can get down to work in earnest. Toads plop over the side every few seconds, sinking to the muddy bottom a few inches below us. Some of the females are trailing ticker tape from their back ends. At last only a mass of tangled spawn is left in the bottom of the boat – an unappetising caviar – so that I decide the only thing to do is to get out of the boat in a suitable spot, tip it up so that all the spawn left over after the party slops into the weedy water. Then we judiciously place a few pairs of toads among it so that sperm still being produced can have a fine old bonanza wriggling round fertilising ova in every direction. Hermione is disappointed she can't see the sperm. I thank heaven we haven't got to give them names too.

It is almost certainly the case that even if these three hundred toads and their accumulated spawn had been abandoned in the moored boat, lost, overheated in the sun, or just forgotten, as so often happens when something else attracts children's attention, it would not have affected the total toad population in the loch. I have not exaggerated; thousands of toads migrate back to breed in its eight benevolent acres. The loss might not have been significant, but an important dimension of being a naturalist is to respect the fellow creatures with which we share our planet. It is a principle dear to me and one I have taught each of my children from an early age, not just because of this, but also to help them build for themselves a rational defence against the sickly animal-sentimentalism that pervades so much of modern children's entertainment. The cuddly animal culture does few favours to children or to the animals it purports to represent.

We moor the boat back at the jetty and I wash the slime from my feet before heading home for tea. I am pleased that the toads are home again after their adventure, hopefully unharmed. Hermione has learned a lot about toads. She has a headful of impressions she will remember for the rest of her life. There is much more she will learn in time, but by then I hope that the wonder and the joy of these adventures will have worked their ageless magic, expanded her imagination and shaped her values. Here, perhaps, is where the beauty lies.

THREE

Dawn and Dusk

If the day and the night are such that you greet them with joy, and life emits a fragrance like flowers and sweet-scented herbs, is more elastic, more starry, more immortal, – that is your success. All nature is your congratulation, and you have cause momentarily to bless yourself . . . The true harvest of my daily life is somewhat as intangible and indescribable as the tints of morning or evening. It is a little stardust caught, a segment of the rainbow which I have clutched.

HENRY DAVID THOREAU, 1817–62

The jackdaw is a crow, and thereby hangs its fate. To some it's like being a garden weed, accursed, scarcely worthy of a proper name, just 'crow'. The sins of its scavenging cousins, the carrion and hooded crows, even the raven's fell reputation in folk lore and fable, are visited upon it for ever. Many people don't know the difference; most couldn't care less. Black, after all, is black. But Hermione knew – perhaps by some sort of home-brewed osmosis – she knew very well. She knew that the jackdaw is different.

One bright spring day some men came with snarling chainsaws to lop down trees along the electricity power lines that traverse our woods. They arrived without the courtesy of notice or permission and seemed surprised when we displayed mild proprietorial indignation. It was, they explained, essential maintenance: in winter snow would build up on branches and cause them to fall on to the lines, plunging us into darkness and gloom. Death, it seemed, might shortly follow. They gave us no choice. They said they had the authority, although from whom we were never quite clear. 'Must you do it in the nesting season?' we asked. They looked blank and cut down a few more. I walked off in disgust.

Listening to this altercation with a child's inquisitiveness fully unfurled, when the men had gone away, my eight-year-old daughter spent her afternoon picking through the tangle of brush and butchered boughs they left in their wake. She found two things of note. The first was a wren's nest jammed in the cleft of an ivy-clad stump that had until a few hours before been a venerable holly tree, much favoured for its Christmas berries. The parent birds were making a fuss. They were chittering and flitting about in a tizzy. Their nest was intact: the chicks, tiny, naked and ugly almost beyond description – more like insect larvae than birds – were apparently none the worse for the rude dismembering of their world. Hermione sat a little way off and watched. In no time at all the hen wren had overcome her proprietorial indignation and returned to her young. To a wren intent upon parental duty, a tree is apparently a tree, whether it has branches or not. Like me, the male continued to tick crossly in a bush.

The second discovery not far away was yet more commotion from a pair of jackdaws. This time the men had crudely truncated a rather distinguished wych elm into the middle reaches of which, many years before, we had nailed a nest box for tawny owls. Sometimes called a chimney because, unlike normal nest boxes, it has no lid or hole, it's simply a plywood cylinder two feet deep, tied on to an upper bough at an oblique angle to resemble a natural hollow in the tree. To be truthful, I think we had forgotten it was there – but anyway, down it came with a crash. It smashed to pieces. The death the men had seemed to foretell had in fact already happened. Inside were four half-fledged jackdaws. Two were squashed. The falling bough had crushed the box, and that was that. The other two must have been thrown clear. They sat in the long grass looking, like the rest of us, but with a good deal more reason, indignant. The parents flew round making a to-do – as might be expected if half one's family had been wiped out.

This find ended up in the boot-room in a cardboard box which both rocked and wheezed intermittently. Looking particularly pleased with herself, Hermione came in with a yoghurt pot full of earthworms. 'Aren't they cute?' she asked, in a tone that made it clear I was expected to agree, as she crammed their muddy wriggling inches into two violently yellow glottal gapes.

'Er, well, quite,' I said warily, not sure whether I was acknowledging the cuteness of the worms or of the two quill-studded, reptilian orphans with bulging bottoms, which squatted, scaly legs askance, in the already fouled box. I knew what was coming next.

'So can I keep them, then?' Dawn and Dusk had arrived.

In fairness to the species, the jackdaw, *Corvus monedula*, is a remarkable bird, which has been much maligned for its habit of blocking chimneys with twigs. Like many of the crow family it is highly intelligent and adaptive. Reference books will tell you that its name is half onomatopoeic, from 'Jack', the bird's commonest call, and 'daw', old English for a simpleton. I think that must be wrong. Firstly, only a simpleton would fail to notice the bird's obvious intelligence (nobody ever called an accomplished thief a simpleton), and secondly, the much more convincing old English (also onomatopoeic) word 'caw' was in common use throughout the Middle Ages for the whole generic coven of black crows: jackdaws, rooks, carrion crows, choughs and ravens. I think it far more likely that the name for this bird was 'Jack-caw' – say it quickly and it hands you jackdaw in a jiffy.

Jackdaws seem to like man; shamelessly they occupy our buildings and follow our activities with scrutiny, never missing a trick. The bird has the cheek of a monkey and the guile of a politician. Dawn and Dusk were comic and fun (Dawn ringed on the leg with a white pigeon's ring and Dusk with a blue – otherwise we would never have been able to tell them apart). As the weeks went by they shed their filaments of down and flaky plume sheaths; the new feathers expanded and stretched like leaves; barbs and barbules linked and layered. Suddenly, one day, with a shake and a rattle, our scruffy chicks were dressed in fine raiment of black and grey, each silken head a dome of feathers masking the quick mind beneath. When, fully fledged, they finally flew to their freedom, I found that I had become attached to their blithe opportunism and their perky companionship. I missed them sorely when they went away. What had started as an indulgence for Hermione – easier to say yes to than deny – became a part of our lives like any other family pet. They had won our respect; our affection followed obediently behind.

There had been many other, less successful orphans before them. Every year the rookery in the horse chestnuts delivers up waifs and strays, most of which are probably doomed to fail. It is always likely to be the weaklings that get blown from the nest, who would never make it to adulthood. By the time they've hit the ground from nearly a hundred feet up they are usually brain-damaged, or broken-winged, or just so shaken-up that they never properly recover. Some have lingered for a week, squawking pathos from the bowels of a box; the suffering of others that were clearly going nowhere I have quietly terminated, and then covered my tracks with white lies.

There have been half-grown thrushes and blackbirds, fluffy leverets, tiny, blind wood mice, mallard ducklings and a procession of pathetically cheeping, weakling bantam chicks. Some have struggled towards adolescence; most have met an early fate by overfeeding or wrong feeding, dehydration or being too hot or too cold, or they've just fizzled out in that remarkable, self-delivered *coup de grâce* wild animals seem to be able to call upon as a last resort. Many times I have risen early and nipped downstairs to peer into a box, fearing the worst – and finding it – stiff, food-soiled corpses, half-gaping beaks, glassy eyes staring nowhere, featherless purple paunches with the rigid chill of stone. Then I have to unleash the grim news – gently, oh, so gently. 'Rookie has gone to a much happier place,' I can hear myself saying, as unconvincing then as now.

It is by such a process that children can learn to accept the inevitability of death, to nurture a little compassion, to attune their formative emotions to the stark reality of nature's wild wheel. The tears, the solemn funerals, the wooden crosses tied with ribbon and the posies of violets and primroses adoringly laid down, are all essential ingredients in the ritual love-potion of childhood. By such perennial enactments children learn to identify with the misfortunes of their hapless charges and through them, perhaps, with those of other human beings and with their own. They are nature's irreplaceable benediction, which hones the character and expands the soul.

Dawn and Dusk were survivors. From the moment the wych elm hit the deck they were destined to make it. I fancy that, had Hermione not rescued them, their attentive parents would have raised them; responding to their insistent wheezing and chirruping cries, finding them in the long grass and dutifully feeding them, just as they would have done in the nest. Even in this wild neck of the Highland woods, where foxes, badgers, wildcats and pine martens prowl throughout the light summer night, I have a feeling that these two would have come through, though young jackdaws have no instinct to hide from predators and have to learn it from their parents. Survival was their thing. It never occurred to whatever buzzed inside their bald and bulging corvine heads that anything else might apply.

It's not so much guilt that I feel about the decades that have slipped by without ever really looking properly at jackdaws (when did I last look *really* closely at a blackbird or a blue tit?), as a renewed gratitude to Dawn and Dusk for forcing themselves and their kind back into my life after an absence of thirty years. An absence since my own first childhood jackdaw, named Rheims after Richard Barham's infamous 'Jackdaw of Rheims', which stole the Cardinal Lord Archbishop's turquoise ring:

> *. . . From his finger he draws*
> *His costly turquoise;*
> *And not thinking at all about little Jackdaws,*
> *Deposits it straight*
> *By the side of his plate,*
> *While the nice little boys on his Eminence wait*
> *Till, when nobody's dreaming of any such thing,*
> *That little Jackdaw hops off with the ring.*

Filching rings apart, the bird has achieved scientific fame with another ring, that of the famous ethologist, Konrad Lorenz, whose flock of tame but free-flying jackdaws at his Austrian home in the 1950s was the inspiration for his work on animal behaviour, and which features so prominently in *King Solomon's Ring*. This extrovert but serious scientist surrounded himself with pet wild animals: greylag geese, ravens, and above all his jackdaws. Although reared in Pavlovian tradition (which expounded the importance of conditioning in animals), Lorenz discovered that 'imprinting' – the now entirely accepted process by which higher animals learn the shape, sound, smell of their own species – was a brief learning phase through which his young jackdaws and geese passed, thereby enabling him to become the focus of their attachment as a surrogate parent. By this process, he incidentally observed, it was possible to make jackdaws into hugely rewarding pets.

The jackdaw is not black. The raven *is* – as black as the ace of spades – as is the rook, and that irrepressible villain, the carrion crow. But the jackdaw stands out for its silken, silver-grey nape – a subtle headscarf that runs behind the eye and over the brow, down to the back of the neck – and, of course, that eye. Like those of the tiger, the eyes of a jackdaw should be a semi-precious stone. They are a startling pearly blue, the luminous blue of a song thrush's egg, with a glint of steel. At the core, pencil sharp, flashes a bead of jet. Nor can the rest of its plumage be written off as black. To see the bird in sunlight, dapper, bustling, head held high at a jaunty walk, reveals iridescent fringes on its back, a blue-green gloss on its wing primaries and a reddish-purple sheen on its secondaries.

Jackdaws are talkative and brilliant mimics. In the Middle Ages they were commonly kept as cage birds and taught to talk, much as the parrot and minah are now, except that the luckless jackdaws had their tongues split in the mistaken belief that this made them better mimics. And they are highly sociable, often flocking in hundreds and mixing freely with rooks, following them as they feed in fields, letting the larger corvid do all the work – turning over cowpats and grubbing in molehills – the jackdaws nipping in at the last moment to snatch a beetle or a juicy leatherjacket.

They learn astonishingly quickly, expanding an innate opportunism into skills the Artful Dodger would have boasted about. They open milk bottles and cartons; in pairs they watch litter bins, waiting for someone to throw something in before they swoop down to inspect, one bird going in while the other keeps guard. They have learned to haul up a string to get at bird feeders full of peanuts. They perch on the backs of sheep, donkeys, horses and cows, scouring their fur for ticks and fleas and, while they're at it, plucking wool for lining their nests.

But it is in flight that they excel above all other crows. They are masters of the air and the wind, and, what is more, they know it and they love it. They play joyously in gales so strong that most birds have scuttled for cover. In gangs they flock and spiral high into the sky. They relish the wind hurling them upwards before flipping over and diving, tumbling and spinning earthwards for sheer delight, pulling out at the last moment with consummate precision and winging up to do it all over again. I have watched them in blasts of over seventy-five miles an hour, gusts in which I can barely stand, performing breathtaking aerobatic displays, often in pairs or apparently coordinated in a whole troupe.

Konrad Lorenz taught his jackdaws to count. He also discovered that they were monogamous and endearingly loyal to their mates for life, or at least until one partner died. He analysed their social behaviour and found they possessed discomfiting human characteristics in their attention to pecking order – they were distinctly snobbish. When one was established as a 'superior' bird the others would suck up to it with affected mannerisms and obsequious deference. After the publication of *King Solomon's Ring* in 1952 a whole generation of country boys grew up wanting pet jackdaws. I was no exception; Rheims was the first of several I raised (before Jack, Jock, Jake and Jill).

But the most endearing jackdaw image I know is that of my old friend Roy Dennis, the celebrated Highland ornithologist, who told me that his tame boyhood jackdaw used to ride perched on the handlebars of his bicycle. When he came to a good downhill slope Roy would pedal frantically to build up maximum speed and then freewheel joyously with wind tears blurring his eyes. The jackdaw thought this was huge sport. It spread its legs for a firm grip, leaned into the wind, raised its tail to horizontal and hugged its wings tight to its sides. On corners it swayed to the right and left, revelling in the slipstream for which, of course, it was brilliantly designed.

*

Every morning from my dressing room window I would call out 'Beow! Beow!' – our nearest imitation of juvenile calls before they had properly mastered their consonants. From their aviary in the courtyard below Dawn and Dusk would answer me in the same ringing cries. I found that I couldn't wait to go down and feed them, a chore for which Hermione and I vied with such rivalry that we had to take it strictly in turns, adjudicated by Lucy. We gave them bread and milk and scraps: bacon rind, rice pudding, porridge, and their favourite, scrambled egg with beetles and woodlice and a teaspoonful of olive oil tipped in for good measure. To begin with they gaped wildly on wobbly necks, their ludicrous, bright yellow clown lips and gullets triggering the parent birds to feed them more. Frantic wing fluttering and an insistent barracking cry accompanied this demanding gape. It is a sound now indelibly imprinted upon my brain. Every May, as I walk round the countryside and through country towns and villages, I often hear this tuneless chorus emanating from hollow trees, chimneys, holes under eaves, crevices in ruins, even once from deep inside an ornamental urn in a graveyard. While filling my car with petrol at the local garage one day, I was absent-mindedly daydreaming when I heard the familiar cries coming from beneath the roof of a corrugated iron shed close by. Before I could stop myself I had uttered a loud, resonating 'Beow! Beow!' The pump attendant emerged from his kiosk looking alarmed, asking if I was all right. I assured him that I was, but for months after that he eyed me suspiciously every time I called in.

As Dawn and Dusk grew they learned to feed themselves and the chirpy calls calmed and then stopped. Their cry evolved to a staccato 'Jack! Jack!' (or 'tjak, tchack,' as it is written in bird books), as well as a wide vocabulary of little whimpering and clicking endearments delivered directly into my ears when they sat on my shoulders. They flew at just over a month old. At first they stayed very close at hand, perching on a nearby bush or tree, looking mildly perplexed by the whole concept of aviation. As we approached with food they would launch themselves at us, landing untidily, occasionally crash landing with a frantic lashing of wings on anything that would bear their weight.

At night we assiduously shut them away in the aviary, fearful that pine martens would make an easy meal of them. The days ticked by. They became bolder and strayed into surrounding trees and on to the roof, chattering to each other in a perpetual happy commentary of our lives. Then, one day at bedtime, Dusky was missing. Hermione and I wandered round and round the gardens calling. We searched roofs and in outbuildings and peered through binoculars at the rooks fidgeting in their high roosts. There was no sign. It was a sorrowful eight-year-old whom Lucy and I coaxed to bed that night.

At first light we were both down to the aviary in our pyjamas to see whether Dusk had come home. Only a lonely Dawn sat there, clearly pleased to see us. For three days Dusk was absent. We feared the worst; hope dwindled. I was convinced we would never see her again. Then one morning just as Hermione was preparing to go to school we heard a mob of wild jackdaws racketing about in a big Douglas fir above the chicken run. 'Beow! Beow!' we called out in unison. Immediately, from high in the branches a crisp 'Jack! Jack!' came echoing back. A second later there she was, skimming down on half-closed wings, a black rag plummeting from a hundred feet up, splitting wide the morning air, seeding the wind with joy, landing with a muffled whirr of vibrating pinions right on to Hermione's shoulder.

I knew very well that those happy days could not last. Dusk had met wild jackdaws, flown the free wind and tumbled in a flock among the tossing treetops. Whatever dark secrets were contained in the chirping and chattering between the two siblings during the next few

days, I find it hard to believe that something of the all-enticing wild was not passed directly to her sister. I began to prepare Hermione for this inevitable parting. I told her how wonderful it was that they would find mates and nest and rear young of their own in two years' time. Wasn't it marvellous, I tried to persuade her, that every time she saw wild jackdaws barrelling across the clouds on a high wind, or caught snatches of their cheerful chatter in the treetops, she would know that Dawn and Dusk were there too?

It weighed mightily. With the joy gone from her own blue eyes she shouldered the reality. She knew it was best for her beloved charges, but she could not hide that she wanted to go with them. She wanted to live their world just as they had lived hers – as I had, all those years before – to extend the bond, hold them close, locked with her in spirit so that she could tumble with them down corridors of sunshot sky. To have been able to magic them down from the heights at will, even for a few brief weeks, calling into the wind and the wind answering her back with its paired shadows of gleaming black delight, had lifted her to the very brink of ecstasy. And when they flew it was as though she was scrawling her own signature across the sky. This was the gift of gods: she was Mowgli and Dr Dolittle; she possessed the magic of Merlin and Gandalf, all her own. To child and adult alike it was unbridled joy. Little wonder she looked so forlorn when, one day not long afterwards, they both lifted from her arm and swept away.

For a week they came and went. We never knew from one day to the next whether they would be there when we called or whether we would be left standing in the courtyard willing their rattle of wings to sweep over the roof. The aviary door was now permanently open, but they never entered. If they came back at all it was to snatch a morsel of favourite scrambled egg, and whisk away again to the wild flock that frolicked and bickered high among the trees. Occasionally they still landed, as from nowhere, on our heads, or, to the profound consternation of complete strangers, on the heads and shoulders of people visiting or making deliveries. One afternoon a local lady (for whom we didn't much care) was trying to sell us political raffle tickets we didn't want to buy. Out of nowhere Dusky swooped down, landed on her shoulder and gave her gold earring a ferocious tweak. With a shriek of alarm the poor woman dived into her car and drove away in vertiginous dudgeon. She has never returned. We giggled helplessly at this timely intervention.

I tried to joke about their inevitable departure. I said they had fallen in with a bad lot from the neighbouring glen (the Highlands is still very clannish). I held out the hope that one of them would bring back a mate, as had happened to Konrad Lorenz, so we would have lots of Dawny and Dusky jackdaws round the house. But Hermione was wiser than that. I knew from her smile and the way she took my hand that she didn't really believe me. Then one day on my way back from Inverness I stopped to buy something in Beauly, four and a half miles from home. There was Dawn, working the litter bins in the square, white ring gleaming, strutting happily about, coquettishly flirting with three wild jackdaws. 'Beow!' I called out, 'Jack! Jack!' She flew straight to me, landed on my arm, touched, and flew off again immediately; she circled round me once and returned to her friends. I tried again but she ignored me. I knew that was it.

'Don't worry about Dawn and Dusk,' I said to Hermione casually when I arrived home. 'They're in Beauly with that gang we often see hanging about the Priory. They're fine,' I added cheerily. For a second her eyes lit up. She smiled at me. 'Thanks, Dad,' she said, but she was looking away.

FOUR

Raymond and the Grass Snake Revisited

. . . for I have learned
To look on nature, not as in the hour
Of thoughtless youth, but hearing oftentimes
The still sad music of humanity . . .

WILLIAM WORDSWORTH, 1770–1850

All children experience unplanned events that pierce the bubble of innocence, that lift their sensibilities out of the sunshine and cast them firmly into the cold shadow of reality; events that come like thunderstorms from behind the mountain, engulfing them and suddenly making everything different, pulling deep focus, raw and unforgiving. For all of us there come moments when lessons are learned; Rubicons are crossed, truths become indelibly logged.

When I was too young to be out fishing on my own, probably only seven or eight, I slipped away from home on a private adventure. I climbed out along the trunk of a gnarled old willow that had fallen into a Somerset farm pond. It was a warm afternoon in one of those endless summers of childhood. There I dangled my hook and worm, string line tied to the end of a stick, for tiny green eels, or minnows – or anything. The shallow edge was muddy and thick with sedges, the water below me stippled with duckweed like a Seurat canvas.

To my delight I spied a moorhen sitting on her soggy floating nest mound anchored to an outer branch of my tree. It was only a few – perhaps fifteen – feet away. The moorhen eyed my every move, but she appeared unperturbed by my presence. I watched her fluff out her skirts and settle herself properly down into the nest. A little while later I saw that she was moving. Something beneath her was absorbing her attention, although I could detect nothing amiss. Her neck rose and her wings ruffled, held half open as if guarding something. Now her breast feathers were fidgeting, pushing up from underneath. I forgot my fishing and watched. She raised herself up to reveal a clutch of newly hatched chicks, black and fluffy, like tiny pom-poms plucked from the hem of an old lady's Sunday shawl.

I was enthralled by this discovery. This was nature the munificent, nature the heartening, the soft and cuddly, nature the utterly delightful to a child. The moorhen began to relax again. She fussed among the chicks with her bill, picking out bits of shell and dropping them over the edge of the nest. She settled once more and riffled her feathers with maternal pride. Her precious brood disappeared into the warm secret of her underskirts. She dozed and I dangled my hook inconsequentially for a little while, glancing every now and again to see how she was getting on. She was an epitome; a metaphor for everything in nature that enchants. Above all, she was mine – a real discovery – more precious to me than a new toy, and I was a willing conscript to her insubstantial aura, which swamped me with a glow of well-being.

A few minutes later my moorhen rose up in alarm. She cried out and scuttered away across the water, feet and wings trailing a channel of fear across the broken scum of pondweed. I couldn't imagine what was wrong. Her brood, still too young to take to the water, sat huddled and exposed. Their tiny red-tipped heads peered helplessly from the black smudge of their bodies. Then I saw it. In cold slow motion a curving ripple approached the nest, inscribing horror into my eyes. The moorhen called frantically from far across the pond. I was transfixed, too petrified to do anything but watch.

A large grass snake, yellow, green and lissom, was winding its way through the stagnant water and the sharp sedge stems to the edge of the nest. In one long, wet and sinuous slither it mounted the rim and coiled its tail around the anchoring twig. A mouth, white and gaping like open scissors, but with pallid, down-curving fangs, flicked forward and engulfed the first fluffy chick. The serpent's head rose and dipped in a convulsive swallow. In an instant the chick was gone. Then another, and a third and a fourth, and the last – all five – in not very much more than as many seconds – to reappear as a unified bulge behind the snake's head, mobile, grotesque and terrifying. The unblinking reptilian eye did not register my frozen form. The blood had drained from my face and my throat was gripped with rigor. I could neither move nor cry out. The silent villain slid back into the water and snaked away through the green slime as quickly as it had come. The nest was empty.

I really don't remember very much more. All conscious thought terminated and memory closed down at the point of the snake's departure, the images too stark, too all-consuming to allow rational thought to proceed. I must have dropped my fishing line and run up the bank, tears blurring my vision so that I stumbled and fell. I was told that I arrived home muddy, grazed and bleeding at the knees, blurting out some ghastly tale the adults were disinclined to believe. Snakes were cruel creatures to be avoided, I was advised. Then someone scolded me, saying that I shouldn't have been there anyway. I was sent off to clean up for tea.

In the space of an afternoon nature bountiful had thrown its arms around my neck and embraced me with warmth and delight, and minutes later nature savage had snatched me up and dashed my innocence to the ground. I would never forget those images; harsh realities permanently tattooed, wired in for good. Their principles would be rigidly applied thereafter. I had learned a lesson. Never again would I permit my effulgent trust to be so ruthlessly gulled by sentiment.

It was inevitable that sooner or later in the bucolic life we lead, something similar would befall Hermione. I hoped it would not be so traumatic, so cold-blooded. I wanted her to arrive at reality gently, with knowledge buffering the blow, a little preparation anaesthetising the first cut. Yet I knew that by understanding the merciless character of nature she might better be able to comprehend the uniquely human concept of cruelty and the purpose of pain.

Comedy defuses anguish. Hermione's first natural disaster when she was five years old had us all laughing so infectiously that eventually, through the tears, she laughed too. It was a situation farce. On a hot spring afternoon she discovered some tadpoles in a ditch, the family Labrador loyally at her side. They were fat and glossy. They shimmied their tails alluringly. Catching them with her hands was easy. The ditch was drying up and they were doomed unless it rained very soon. She put them in a plastic mixing bowl of clean water – all good childhood stuff – but she overlooked the dog. The tadpoles ceased to struggle, hovering contentedly in this cooling release from their rapidly overheating natal ditch. They looked like an expensive hors d'oeuvres, like some exotic seafood awaiting the aspic. In the way that a child instantly confers affection, Hermione loved them all. The May morning had grown hot. She went indoors for a drink. The dog took one too, and downed the lot. Screams rent the air. We ran out to rescue her from something that must surely be life-threatening. Her face was contorted with anguish, tears flooded copiously down her crumpled cheeks. There was no blood, no obvious injury. Through her wailing desperation she pointed to the mixing bowl. The Labrador stood at her side, loyally wagging his tail.

Later there would be other dramas; the picture fills slowly. By virtue of constant exposure she was learning fast. One June evening, as the light began to fall, we sat together in a window and watched a pine marten doing the rounds of the nest boxes on the trees and buildings around our home. I explained to her that this was a trick the marten would have learned by chance experience. He would have noticed birds going in and out, so he climbed to exploit the nest, just as he would a wild nest site. After a few successful raids on boxes, delivering up a meal of eggs or chicks, or even an unfortunate adult robin or flycatcher caught sitting, he would have begun to associate the box shape with success. It was then only a matter of time before it became rote – the automatic inspection of all box-shaped structures.

The observation lasted only a few minutes, removed by distance and the pane of glass. We heard no crunching of soft fledgling bones, saw no blood, failed to hear the frantic alarums of the adult robins.

But the lesson took root. We talked it through. She learned that these birds were prey and the marten a predator. That the golden eagle snatched pine martens when it caught them out in open country, and the martens ate robins . . . which ate spiders . . . which ate ladybirds . . . which ate aphids . . . which sucked the juices from plants . . . that all nature is a percussion of piercing, stabbing, tearing and ripping, a concerto of unabated anguish. We went to the robin's nest box. It was empty. Cleaned out.

'Will she nest again?'

'Yes,' I reassured my daughter. 'She will find a better place and try all over again.' (To foil the martens we now construct our bird boxes out of odd shapes, bits of old log with natural holes, other such *trompe l'oeil*, leaving the square ones in place but full of sawdust so that the birds are unlikely to use them.)

The 'cruel' word did not emerge. If it was near the surface there was no sign of it. I was pleased; I have such difficulty with our contorted notion of cruelty. The definition I favour is 'disposed to inflict pain', to which I would only suffix the qualification, 'humans'. Our pine marten had no pain in mind; he was disposed to feed himself, that was all, as was the grass snake. The fact of pain, both directly to the helpless fledglings and indirectly to their distressed parents, was real enough. But the disposition of that pain – the predilection, the premeditation of pain – was wholly absent. So was cruelty. Nature is innocent; it neither indulges nor acknowledges cruelty at any level.

We humans, Hermione and I, insulated behind our windowpane, were predisposed to see what the pine marten would do, pain or no pain. We watched with more than a little interest – not to say eagerly. The pine marten is sleek, agile and beautiful, and it was, at least until very recently, a rare mammal in Britain – something well worth seeing. I knew for certain that the predatory process would involve pain and distress. I could have stopped it. We could have opened the window and shouted. Not only would the pine marten have vanished in a streak of orange and mocha fur, but also the robins would still have had chicks. But I was disposed not to interfere with the infliction of pain. Did that make my actions cruel?

Last winter a whooper swan flew into the high-voltage electricity lines that traverse our glen. Every winter these cables take a sorry toll of migratory swans. This one fell to the ground in a swampy field. Its wing was broken. The sharp bone protruded pale and bloody through the layers of white feathers like a broken timber. Its leg was badly twisted, probably broken too. It shuffled helplessly into a patch of rushes where it lay, blinking out at me the pathos of its own inevitable destruction. There is no effective cure for a swan's shattered wing – certainly none that would allow it to fly free and to return to the Arctic to breed. The leg simply made things worse. At this point no cruelty had occurred. An unhappy accident had befallen an unlucky swan. It was in great pain, I am in no doubt about that, although I am glad that it lacked the cerebral capacity to understand either what had befallen it or the terminal implications of its injuries. Its sad, oriental face tilted and pivoted on its elegant neck. All it knew was the fear of my presence and that it could not fly away. I gazed at its slightly gaping, black and yellow fluted bill. Clouds of dread drifted through its liquid eye, making me wince. Remorse flooded in. I wanted to reverse the clock and see it fly again in chevron formation with its seven fellow swans; my spirit longed for the rhythmic whoop of its wings and its wild bugling, summoning the beauty of the morning.

Nature would recycle this swan. Not quickly – it would have starved, or been torn to bits by foxes when it was very weak – but it

could have lived for days. I preferred to kill it than continue to live, even for a day, with the knowledge of its long, painful demise. In less than a second it was dead, its thin skull crushed beneath my heel.

Twice I was disposed to inflict pain: the gasp of its departure and the emptiness of my numbed brain. But it was quick and certainly as stress-free as any natural death that might have overtaken it.

And what if I had ducked it? Walked away, pretended I hadn't seen its ignominious tumble from the sky; blocked it out to save my own pain, even though I knew full well what prolonged suffering that action would cause? Would that have been cruel too: the compound cruelties of omission and indifference? Is the quality of mercy really unstrained?

We live by instilled standards of conduct. Human activities are full of inconsistency, not to say humbug. A person who would fervently decry the apparently cruel sport of a neighbour is guilty of failing to check whether the rabbit or the hedgehog he hits with his car is dead or alive. Another who is shocked by his small son pulling the wings off a fly thinks nothing of wounding and maiming many birds during a day's shooting. A vegetarian who can't bear the thought of calves and lambs being killed, pursues the death of a rabbit in their garden by any possible means. Those who profess to love birds and wildlife, who join clubs and societies for their protection, donate money and sport ties and badges proclaiming their allegiance, often think nothing of spraying the garden with toxic weed killers and pesticides, not bothering to read the warnings on the packets. Others care passionately about the clubbing of seal pups on the Newfoundland pack ice, while setting mousetraps in the larder expressly to deliver up violent and often torturous death. A fox being killed by hounds is offensive to many who could not care less how a rat dies – both warm-blooded mammals, both intelligent, both pests.

We are happy to love killers when they suit our romantic perceptions; we are affronted by others that don't. Poets and bird lovers idolise and sentimentalise the song of the thrush, choosing not to acknowledge the killing that sustains it; a sparrowhawk in wild innocence snatching a chaffinch from a bird table is damned as cruel and savage. All these inconsistencies seem to me to prove that cruelty is an abstract, and one that, in the final analysis, remains critically a matter for subjective conscience.

Aged nine, Hermione knows little of this. In so far as it is possible to foster the development of a conscience, I do my best. 'Be thoughtful,' my grandfather said to me when I was her age, 'and have respect for the creature, whatever it is.' The passing down of such principled wisdom provides a glow, however diffuse it may be. But it shed precious little light on the dilemma of Raymond.

Our property has honey fungus, *Armillaria mellea*, a devastating parasite for trees and shrubs. It infects roots and butts, invading the lignum and turning it to mush, eventually extending high into the host plant, killing it with a white and fibrous rot (which, most curiously, is luminous in total darkness). It is endemic. There is nothing we can do about it, no remedy, no escape. It's an unseen enemy, like living with guerrilla warfare; we wait to discover where it will strike next. Last May a birch tree fell unexpectedly across the drive. The butt and the lateral roots – the anchors that keep a tree upright – were heavily infected; I could push my finger into the timber with ease. I was amazed that the tree had stood so long. We cleared the brash and began to cut the trunk into discs.

Determined never to miss out on such excitements, Hermione came with me to inspect the damage. She found a hole in the bole at what would have been fifteen feet from the ground. She peered in. Its fungally interior revealed a nest about two feet into the darkness. 'Daddy, Daddy, come and look!' Sure enough, four ugly hatchlings lay sprawled in the nest of feathers, grass and sheep's wool. They were starlings, *Sturnus vulgaris*, that ubiquitous and irrepressible knave, the epitome of bird impudence, the hangers-on to human habitation and activities: chirpy, omnivorous, characterful and scorned. These were tiny and naked – mere mouthfuls the pine martens had failed to find. Their domed heads and bulbous unopened eyes wobbled feebly on stringy necks and their ludicrous yellow beaks gaped and cheeped at us with all the bewildering naïvety of blind instinct. 'Let's keep them,' she begged.

My own blind instincts said walk away. Pretend you haven't seen them. They are far too young to survive; they're a headache and a mess I can do without. Aren't there enough starlings in the world without these four? 'Be thoughtful and have respect for the creature, whatever it is,' had returned to mock me – irony hovering like a bluebottle. I gave in. 'OK, OK,' I said with a sigh. 'As long as you catch the bugs.'

From the very beginning the starlings were a huge success. They grew rapidly, eating virtually anything: beetles, worms, woodlice, scrambled egg, porridge, grated cheese, mashed potato, tinned cat food, even prawns. Hermione was faultlessly attentive. She rose early to catch fresh bugs and loyally replaced the starlings' fouled sawdust before she went off to school. Upon her return her first goal was the cardboard box beside the Aga. They knew her voice and cheeped furiously to be fed. Out came the tweezers and hapless beetles disappeared into four bright yellow ravening maws, all straining to out-yell each other. Even Raymond, the pathetic runt of the brood, did his best to compete.

We never knew the full extent of Raymond's problem. It was surely genetic: a deficiency locked into his deepest coding. He was half the size of the others, shoved aside all the time. I am sure that in the birch tree he wouldn't have made it past the first week. The others were so strong and insistent that he wouldn't have stood a chance. As he weakened they would have sat on him, smothering him and pinning him down, so that when the parents arrived at the nest with food he would never even have got to raise his head. Once dead the adult birds would have hauled him out and dropped him to the ground – so much dross – nature systematically rejecting and recycling the unfit.

But Hermione was having none of this; she had set her mind upon reversing the process. Raymond was singled out for intensive care. Beetles were crushed and dismembered so that he could better cope with them; caterpillars of scrambled egg were delicately oozed into his gullet through a plastic syringe. She would remove him from the box altogether and sit him on her knees while she lovingly coaxed him to take the tenderest morsels. He responded. He remained small and weak; he made less noise, but Raymond survived. He grew stubby feathers and ejaculated pleasingly normal-looking globules of excrement from the pursed lips of a cloaca instinctively pointed away from the nest. This endeavour was so particular, so purposefully performed with spread legs and straining neck, the whole of his tiny body angled and directed with such innate determination that it seemed as if by some supreme act of will he could also eject the malevolent gene that sought to deny him life. Hermione whisked away these black-tipped, chalky-white mucus bubbles every bit as attentively as the adult starlings would have done.

For two weeks the fostering of the rowdy three and Raymond continued apace. They grew cheeky and comic. With eyes now wide open they watched every movement in the kitchen, nosily peering out over the rim of the box. Raymond, it seemed, was almost back on track. Downy plumules began to flower from their papery buds; their bills elongated and sharpened, leaving the vulgar yellow flanges of infancy still bulging at the corners of their mouths like an over-sized glove creased between thumb and forefinger. They were no longer the helpless reptilian caricatures we had rescued from the stricken birch. They were on the point of becoming proper little birds.

Three did. Instead of squatting, they now perched; they began to preen; feathers expanded, soft and slate-grey; they stretched their wings and scratched behind their ears with a clawed foot. But Raymond had stuck. He suddenly stopped developing. His metamorphosis seemed to be terminally incomplete, neotenous, a larva for life. Something was wrong. Heroically, his cells had struggled to evolve, egged on by Hermione, but their most arcane inner workings had finally let him down. Raymond had ground to a halt. He was a bud that never opened, a fern locked in crozier phase, a tadpole unable to shed its tail and break free from the pond. His feathers remained stubby and he had whole bare patches on his wings and breast where none grew. To make matters worse, his eyes began to close. He seemed sleepy all the time, as if his batteries were low, constantly dipping into standby mode. If we nudged him they opened and flared with life for a moment, then the membranous lids slid across and shut him in again. His head began to droop; not just sinking to rest as a bored dog opts for sleep, but with a sinister downward curving of his head so that his bill ended up between his little clawed feet, awkwardly snagging in the sawdust.

We watched in dismay. The signals were grim. I had known all along that he was out of the race. I had often thought of telling Herminone; had braced myself to explain the rough old rules of natural selection, but I had put it off – as one does. I prayed that no water serpent would slither in and engulf her precious chicks, especially not Raymond. I had elected not to cast the shadow of doom over what had been such a happy, sunlit affair. Now it was too late. Raymond was her favourite and he was closing down; just fizzling out.

I knew that she knew. I saw her lift him out and cup his insubstantial frame in her hands, raising his beak to her lips as if to breathe her own life into his failing breast. I heard him cheep feebly and fidget his bill with the sound so that he seemed to be speaking to her, a private exchange between parent and child. Then I saw him sink into sleep in her hands. I wished that he would die there, right then, so that she could draw grace from this final act of oblation. But it was not to be. Raymond clung to his wisp of life a little longer. I realised that he had arrived at a plateau beyond which he couldn't progress. His tiny crippled physiology had extended itself to the limit. Each morsel of food so lovingly administered allowed only a brief perpetuation of a life teetering on the brink. If the food ceased even for a few hours, he would topple into the abyss.

I dreaded the rigid corpse of the dawn. It seemed to me to be too harsh, the serpent returning from the stagnant pond: a rejection too soon in the school of nature's emancipation. I, too, had parental duties. Perhaps I should intervene – I had done it before. I had been disposed to avoid the pain, and had lied to salve it. No longer did I know where the greater cruelty lay. I was disposed to inflict the pain but ill disposed to suffer an inevitable grief. The swan seemed straightforward now – I had never really doubted where my duty to that poor broken creature lay. But this was different, this was a conflict at the very heart of sentience. My brain was burdened with reason and my heart overloaded with pleadings.

Sometimes light beams through. Suddenly I glimpsed a deeper tract of the no-nonsense wisdom of nature. The pine marten suffered no qualms. Hermione had understood that what she witnessed that day was right. Pine martens eat robins – no fuss, no excuses or laboured justifications, no shilly-shallying, no sharply drawn breath. Suffering is written in for robins, just as it is for all wild creatures. It is life's unavoidable cross: a given, so much data. It's part of the process, part of the great swirling maelstrom of nutrients of which we are all tiny, bewildered fragments. We shut ourselves out at our peril. Do we really know better just because we choose to see what floats to the surface, the birdsong and the sunlight playing in the dappled leaves of summer? Is the urbane human concoction of cruelty a bonus, adding to the stature of humanity, or does it only fog the issues, blur all the edges so that we fall back upon sentiment, further clouding our already distorted vision? Perhaps, after all, cruelty is a smokescreen – I suspect the truth is that most of us could not bear a clear vision of the natural world.

Hermione came to me in the garden. I was quietly pruning a cherry tree that had been damaged by unseasonable wind and rain. Its shed blossom lay around me on the grass. I could see from her gradual approach and weighted countenance that she had something to say. 'I've been thinking about Raymond,' she began. The blossom seemed to turn to snow. I felt like Oscar Wilde's selfish giant standing in a winter of my own creation.

'It's not really fair on the other starlings, is it?'

'What isn't?' I asked, warily.

'Well, if he's got a genetic problem he shouldn't breed, should he?'

I took a deep breath. 'Even if he survives long enough to breed,' I told her carefully, 'I'm afraid the other stronger starlings won't let him. They'll just push him aside, like his brothers and sisters do now.'

There was a pause and then she said, 'I think we've done our best, Daddy, don't you?'

FIVE

The October Roaring

We are Nature, long have we been absent, but now we return,
We become plants, trunks, foliage, roots, bark,
We are bedded in the ground, we are rocks,
We are oaks, we grow in openings side by side.

WALT WHITMAN, 1819–92

It is the last day of September and autumn is in the air – and across the hills, and in the woods, and down at the river fields. It has been setting out its stall for some time. We are still three weeks away from the famous Highland gilding, nature's Midas touch, which, for a brief and ecstatic moment in the year, drifts through the birch-woods leaving in its wake a burnished wonderland so mesmerising that drivers have been known to fail to take corners on this glen's single-track roads, ending up in the ditch. A hint of yellow is creeping into some trees – a few leaves here and there – and those of the gean, the Highland name for the wild cherry, are blushing at the edges like ripening apples. But the really urgent signals we are receiving so far this year have little to do with colour or trees.

Autumn comes in many movements. We should never forget that it is only a prelude; it's an overture building to a much greater oratorical grandeur, to the all-enveloping solemnity of winter. By September the robin's bell has mysteriously cracked; the old familiar warble that heartened us all summer has developed a rattle, a metallic edge, which now seems to contain a hint of presage, even a warning. It is our last daytime songster, the robin, all the others are silent, or have taken the hint and fled. The blackbird, whose rapturous flute is the most melodic of all our common garden birds, has taken the huff, refusing to play. Although they stay with us all winter, now we only see their busy leaf-flicking bug searches beneath bushes, or hear their hysterical alarums when a sparrowhawk streaks through.

A fungal fiesta has fairy-taled the woods: glossy ceps like glazed buns; fluted chanterelles, ocherous and prized; dangerous fly agarics, red-capped with white warts – a little too like children's make- believe toadstools for comfort; the blusher, a pink-fleshed, innocuous-looking but savagely poisonous agaric that lurks enticingly under the pines; and the hordes of unidentified mini-parasols, delicate and pallid, that come and go without any applause, except perhaps from greedy red squirrels and wood mice hurrying to lay down fat for the winter. And there are the mists that hang over the river in the early morning, lingering long after the sun is up, especially after cloudless nights of opalescent moonlight have ushered in the first frosts, white-pluming the breath from my old horse and the Highland cattle ruminating in the pasture. Across the hillsides a touch of umber has swept through the bracken. All these and many more signals are constantly pressing in around us as the dying summer fades away.

Soon, probably in the night when the land is quiet, I shall hear geese high overhead. They have left their summer breeding grounds in the Arctic and are heading south. They'll come high, a thousand feet above us, and in vee formation: long skeins of haggling, bickering greylags or pinkfoots that will inevitably draw me to the window to catch their stirring song. It's an irrational response because I know very well they will be with us all winter – they gather in tens of thousands on the tidal mudflats of the Moray Firth – and soon we shall hear them every day. In a few weeks the winter skies will be constantly threaded with their wavering lines, but it is six long months since I heard that sound and I find that I am always hungry for them again, as one longs to see an old friend. The vanguard, the first to arrive, seem to possess that special indefinable quality of seasonality; the shift in birdsong, there one minute, gone the next, or the long evening shadows that our latitude affords us from the daily lowering sun. These are nature's subtle abstracts that harmonise to make life in the Highlands the sensuous experience it is.

But it is none of these that have drawn me and my daughter out into the night, when most ten-year-old children should be heading for bed. We are tramping up through the darkening pinewood. We're well wrapped: coats and scarves and gloves, as our breath mists against our tingling cheeks. The moist forest air is seasoned with the delectable potting-shed aromas of mould and latent fertility. On the edge of our torch-beams tall pine trunks loom past us like furtive presences. We are heading uphill, up to the forest edge where the moor spills out in front of us like a great heather lake dotted with islands of luminous bracken, undulating away from us, high above the valley, reaching far into the dark mountains beyond; hills that ring us in, glowering and resentful. It's here, on the forest edge, that we can find a mossy cushion beneath a pine tree and settle down, not just to hear, but to *feel* the October roaring.

We have done this every year since Hermione was six. It is part family tradition, part ritual and partly the indulgence of a naturalist father who has striven to share with all his children the excitements of the wild as he first knew them long ago. Ritual, because the roaring is of the essence of the Highlands, in its blood, so to speak, and therefore somehow requires us to celebrate and perpetuate it, as does the native Gaelic. *An dàmhair* is their word for both the month of October and for the red-deer rut; in English, just 'the roaring' – the term commonly used in these glens to invoke the unique qualities of this delectable moment in the turning year.

Red-deer stags and hinds live apart. For most of the year they form large groups of like gender, often occupying hill ground well separated from each other, on opposite sides of the mountain or the glen. All summer the stags can be seen dotted through the heather, quietly grazing and growing their new bone antlers in silent preparation for the roaring. The old stags with noble, widely branched heads of ten or more points know it well; they lie gently ruminating and chewing with a look of savoir-faire on their haughty faces that seems to say, 'You'll see – just wait and see.' The youngsters don't; they haven't yet worked out quite what their antlers are for, or even what this whole masculinity business is all about. They skitter about, teenagers prancing a safe distance away from their betters, jousting and fooling idly amongst themselves, gently testing their hardening weapons, posturing, buzzed by the rising tide of testosterone, but blithely ignorant of its consequences.

Across the mountain the hinds are busy giving birth and leading their wobbly calves into high corries, as far away from bothersome flies and human disturbance as they can get. Wary of eagles, they eye the crags and the scudding clouds. Remembering the wolf, their long angular faces are ever alert, black noses twitching, ears rotating like antennae, dark eyes filled with the primeval pensiveness of the hunted.

Summer slides away. By late September the stag herds have broken up. The big boys have risen from their beds, shaken and stretched and quietly moved off. Through the lengthening nights they have trailed the heather paths of their cervine ancestors in these glens, following the wind, crossing mountain and burn, picking over scree and trailing through bog, running the forestry fences, slowly and purposefully making their way into the hinds' territory and their genetic future. They stop eating; food is a distraction. They are slimming down to reveal their prime; purpose speaks from their quickening pace, proud heads held high. The velvet coating in which the year's antlers have grown has long since dried and peeled away, leaving glossy ribbed horn as dark as mahogany. Whatever surging hormones and their complement of genes can achieve is now done, immutable for a full year, fixed in a hard, heraldic array of branched points and spikes, an achievement – an armorial bearing whose noble pedigree is arrayed for all the world to view.

What happens next is down to stamina, nerve, luck and sheer bodyweight. The hind herd is about to be carved up by bullying male domination. A heavy stag can hold about twenty to twenty-five hinds together in a harem. He will usher them away from the others, working in little rushes like a sheep dog, tireless and persistent, herding them off to a stand of ground he can defend from the other stags. Often this is in a hollow or on the moorland edge of woodland. He impresses his females by stature and bluster – spread of antlers, shaggy black mane, and, above all, voice. Every few minutes he will throw back his fine head, antlers passing either side of his shoulders, and his gaping jaw will angle to the corries as he roars his intentions to the whole glen.

It is guttural, unforgettable and humbling, immediately cutting the listener down to size. It's one of those prehistoric utterances nature has honed for maximum effect; a sound that belongs to the mountains, forests and moors and to our hunter origins, obscure and instinctive. It has nothing whatsoever to do with man's modern, soulless world of froth and triviality. To the other deer, both stags and hinds, it is both challenging and threatening, partly bluff and a lot of bravado – it resonates with purposeful swagger. It sends a shiver down your spine, stops you in your tracks, cutting to the core of your nervous system and forcing you to pay attention. It comes again, louder than before, a long, moaning bellow, rising and building in intensity and power – an outpouring of all those months of waiting and growing – until it lifts to its monarchic climax, holds for a straining moment of pure machismo, before tailing off into a grunting, ill-tempered diminuendo of steaming breath and bad grace.

Never doubting himself, the stag strolls to a boggy hollow and thrashes it to mush with his antlers and front hooves, rending the water-filled sphagnum into a deep wallow. He immerses his whole body in peaty ooze, emerging after a few minutes like some terrible prehistoric beast, lumpy, dripping, glistening, shreds of sphagnum trailing from his antlers. Then he roars again.

It isn't dark. It was back in the wood, but out here the night is soft and glowing. A hemisphere of harvest moon drifts in and out of fine cloud; far away the hard black line of the mountain horizon comes and goes. Above this line and below the clouds a ring of stars seems to signal to us in an excited Morse code of electricity. A tawny owl has skirted low along the edge of the forest, logged our unconcealable infrared images and veered silently away. We can hear a car far below us in the glen; its lights dim and flare haphazardly along the winding riverside road.

We don't have long to wait. A stag is out there in the darkness, about half a mile away, roaring his potent eminence into the night. We have no need to be quiet so we chat amiably. We play a game of guessing how long it will be between roars and whether our stag is static or on the move, whether he is in the forest or out on the moor. Hermione has a new watch with a timer and a tiny glowing lamp built in, so she presses the button to set the stopwatch. 'Ten minutes,' I suggest. She opts for fourteen and a half. Only three minutes later another roar echoes to us across the heather. It seems closer, but from another direction. Excitedly she presses the button again. 'That's not the same stag,' I whisper. We wait again. This time it is much longer. Fourteen, fifteen, sixteen minutes pass.

'Do you think it's gone away?' she whispers. As the words leave her lips a roar is building. It breaks out into the night and sounds much, much closer. It seems to be in the forest behind us. It comes and comes, engulfing us, and the pines and the hills and the moon and the clouds, wrapping us round and then tailing away in a throaty grunting moan of tetchy bravura. Hermione is gripping my hand – tight.

Four and a half minutes this time, but the sound comes not from behind us – this stag must be way out on the moor, definitely coming nearer. We stop trying to guess; the game isn't working the way we expected. What's happening here is a challenge – a gauntlet hurled down. One of these stags has hinds, the other hasn't. We don't know which. One is likely to be static, sticking close to his females in case another challenger nips in and carves up his harem; the other is an opportunist, a usurper on the make and take. The roar seems stronger now, more insistent and focused; we don't have to wait long for the response. Our stag is in the wood – really quite close – perhaps only about a hundred yards away to our right. He bellows rancorously. The moon ducks into denser cloud as if taking cover, and darkness swallows us up. There is a long silence. When the moon re-emerges I can feel the tension slacken in Hermione's grip. The darkness made everything seem closer, more pressing.

Challenges echo back and forward for the next twenty minutes. Sometimes we think they are closer, and then they confuse us by seeming much further away. 'Have you heard enough?' I whisper. 'We could begin to head back.'

'OK,' she agrees. Even as I am drawing in my feet to get up, another roar bursts upon us from behind. We almost duck with fright, it is so close. It seems to be just behind our little clump of pines, in there, only a few tens of yards away now, consuming the whole wood in its urgent, angry, hormone-surging vitality. This close it is a sound that dismisses all reason, brooks no argument. We freeze. We can hear him moving now; he is pawing the ground with a hoof. There is a thrashing, snapping sound as he beats up a small pine sapling with his antlers. Then the challenger roars again, but in front of us, just out there in the darkness, closing in every time, heading our way.

Quite by chance we have positioned ourselves in the middle of this ancient, wild Highland affray – *an dàmhair* – something it would be difficult to achieve except by accident. The show is coming to us, although we can see virtually nothing of it. It's too dark to make out any feature other than the skyline and the branches of our pines etched against the stars. But sight is not necessary. The drama works just as effectively as a radio play, yet far more so because we are in its midst, with the night air caressing our faces, breathing the rancid effluvium of pungent deer musk that floats across us like a black mood.

The stags roar again. The challenger is heading in from the moor, only about fifty yards away now. We can picture him as starkly and vividly as though it were broad daylight. His hard cleaves thud across the peat and the woody heather stems obediently swish to his forceful strides. Our stag, the defender of hinds, is still in the trees, still behind us, somewhere off to our right, but audible too, as is his harem. We can hear the stamp of hooves coming in little rushes as he attempts to hold his hinds together, ringing them round, herding, bunching them against the threat of breaking away. The woods are suddenly alive, full of movement and excitement; the sound of deer is all round us.

We are thankful there is no wind or we should have been discovered long ago and the stags and hinds would have melted away in a flash, streaming through the trees and across the moor, rushing headlong to safety from man the great predator, man the heart-stopper, the party-spoiler, man the dread. Not even the rampant testosterone of these fine beasts would have dared defy the vortex of fear that we humans instantly inspire. But we are lucky. Our scent remains our secret, wrapped with us in the dark, two huddled figures tucked up together at the foot of the pine.

It is suddenly too much for our stag. Something inside him has burst like a dam; choler and indignation flood his veins, surging free and turning him angrily away from his hinds. He can stand the challenge no longer; out he comes, high stepping, as proud as a jousting knight. Leaving his hinds among the trees he powers forward with head held high. We can hear the swift rushing of his legs through the heather, a surging wave of sound, direct and purposeful, out to defend his honour. We raise our binoculars to scour the darkness. We can just make out a shadowy shape – then it is two. There is a jarring crash of antlers meeting and locking, two proud heads clashing together with all the combined forces of our largest land mammal's wild and unchecked rage.

For the next few minutes – it could have been five, more, even ten – we are rapt, hearts pounding, holding our breath, unable to speak or move. The stags come at each other again and again. The forest rings with the hard, echoing clatter of bare bone clinching, shunting and freeing – only for seconds – before crashing together again. The whole moor seems to heave with the energy being churned out against one another, now turning this way, now back again, rending the peaty soil, the heather ripping from its roots. When they part and back off, mouths open, tongues lolling, shaggy flanks heaving, we, too, gasp quick lungfuls of oxygen before wincing visibly as they slam together again. They seem to be evenly matched, these vying, would-be-monarchs of our glen – although we can see neither beast in anything but the dimmest outline, it is clear that they are both big strong stags. I am fairly sure that one is a youngster snapping at the heels of an older patriarch, their shared genes vying within just as fiercely as their antlers clash without.

We cannot witness the end of the duel. The moon dips us into darkness again and we lose all sight of them – not that it matters, the sounds and the rank pungence enveloping us delivers all the urgency and the drama of these last moments as though we can see every move. This ancient animal odour seems as old as the hills themselves, a smell of the animal age before man purged himself of wildness. It is a rich scent that sends shivers down our spines, holding us close, intimate and pressing. At one last, momentous heave the anchoring back legs of the challenger have slipped and buckled. Antlers instantly break free and the master has lunged his advantage home so that the usurper turns and runs for his life. We hear them gallop away into the darkness, the short sharp chase of ultimate humiliation. We know it is all over. There is a long pause – one of those deafening, burdened silences only the night can fashion. Then the roar comes, well away from us, rising out of the moor like something volcanic, a long, triumphant, swirling roar of the victor's contempt for the vanquished.

We sit still. It seems irreverent just to get up and go, whether there is to be an encore or not. All that is out there is empty darkness. No owl hoots; no hinds shuffle nervously in the forest behind us – nothing. The minutes tick by. 'Where's he gone?' Hermione whispers.

'I don't know,' I answer quietly, shaking my head. 'I don't know what's happened.' I had expected him to come running back to round up his hinds and lead them off to his chosen rutting stand where he could hold them secure once again. With the deer and the moon gone, the mountains and the forest seem suddenly distant and aloof, no longer involving us in their affairs, holding their secrets close, as if there were some rule operating out here: wildness for the wild alone. 'I think we should go home.'

'Can I try a roar?' she asks, getting to her feet.

'Of course,' I answer, smiling. We have tried this game before in previous autumns, pitting our feeble voices into the night to see if we could elicit a response. We never could. She cups her hands to her mouth and utters a small, childish, almost dog-like sound, more of a howl than a roar, but delivered with more vigour and passion than I expected of her tender years. I press my knuckles into the soft forest moss to heave myself on to my feet to go home. As I do so, I am chuckling.

The night air is rent asunder like an express train emerging from a tunnel. It is behind us, beside us, all round us. The roar comes and comes. I pull Hermione to me and hold her tight. On it comes, pushing out grand, irresistible, elemental power like a great wind. We are breathing his hot, foetid breath. Acidic musk swamps us, catching in our throats. Instinctively we cower. We know neither where he is, nor whether he knows of our human presence, only that he is close – very close – raging and fired up with the success of his victory, hell-bent upon evicting whatever beast had the temerity to raise its voice within his domain. He's too close – I am worried. The game has taken a dangerous turn. We have witnessed his power; we know only too well how angry he is.

I jump to my feet and shout, 'Hallooooo-ooo! Hagh! Hagh! Aaaarrgh!' I wave my arms and my torch wildly, still shouting, striving to be as human and visible as I can, hurling light beams into the darkness in all directions, praying that man-dread will overcome his valour and the brutish pride that unquestionably swells within his heaving breast. It does. We hear him turn and run. We hear him crash away through the pines in a rumble of galloping hooves on the dry forest floor. In a second he is gone. Far off we can hear the hinds running too. Only then do we laugh. We hug in the empty darkness, laughing that her little roar has worked, bringing him with all his huff and puff to only a few feet away.

'Now that was a roar,' I tell her, as we turn for home.

SIX

Treshnish and the Pippin

I go to nature to be soothed
And healed,
And to have my senses
Put in tune once more.

JOHN BURROUGHS, 1837–1921

Churlish October skies skulked overhead as we waited on the north pier at Oban. Inshore trawlers and colourful crab and prawn boats with evocative names – *Silver Sprite, Dawn Treader, Maid of the Isles* and *Golden Endeavour* – lay hugging each other like sleeping dogs; from another pier a huge black and red Cal-Mac ferry to the Western Isles slid away like a bit of the town breaking free. Two fishermen lurched out of the hotel bar behind us, blaring pop music on to the rain-wet quay. One man cursed the weather, spat noisily and bade his companion extravagant farewells. Then they staggered off, arm in arm, down the narrow street cluttered with fish-and-chip papers and polystyrene cartons.

We were excited and apprehensive, awaiting a boat of our own to pick us up. Out in the Sound of Mull the sea appeared suspiciously calm. A pale sun groped through the lumpy clouds, periodically daubing the sea Prussian-blue and elephant-grey in great clumsy patches, stabbed on as if an angry painter was rushing to catch their constantly shifting effect. The fishermen had good reason to curse. The dying days of September and the first week of October are notorious for equinoctial gales on the west coast of Scotland; gales that appear suddenly on scouring Atlantic westerlies; gales that career wildly through the islands and lash the mainland with furious, stinging squalls. Lulls between these oceanic riots are often bright and magnificent; hours pierced by dazzling sun that picks out islands and headlands, random oases of spot-lit sea and white surf crashing on to rocks and skerries, while unforgiving rainsqualls lay siege to distant, brooding hills.

Hermione looped her arms over the harbour rail and hung like a scarecrow; herring gulls bickered fractiously from a warehouse tin roof. Out in the bay, Bobby Archibald, our skipper, was hauling up his mooring while the *Pippin*'s engines nudged her gently into the tide. Gleaming white with varnished teak deck and bulwarks, the *Pippin* was ours for four days. She was fifty-six feet and thirty-five tons with twin diesels, a Fleur de Lys motor cruiser for a private adventure into these petulant seas at a moment in the year when most sea-going boats are securely moored in harbour and safe anchorages. We were chancing our arm – taking the very risk nature has reckoned and concluded that man is neither brave enough nor daft enough to chance. We were heading for a grey-seal breeding-colony, twenty-five miles out into the Atlantic, beyond the furthest-flung headland of the Isle of Mull. Keyed up like adventurers from an earlier age of exploration, we were bound for the remote and uninhabited, straggling archipelago of the Treshnish Isles.

A very long time ago, probably as long ago as ten thousand years, men first arrived in the Highlands in a period now known as Mesolithic. They were Middle Stone Age and, we must assume from the scant evidence of their shell middens and cave hearths, they strode about in clothes made from animal skins, bearing clubs, spears and bows and arrows – the unassailable stereotype – fishing, hunting and gathering in organised gangs. They had coracles and dugout canoes and made forays up rivers into the densely wooded interior and the mountains, into a natural environment of climax forest so pristine after the gradual retreat of the ice sheet and the glaciers some two thousand years before, that we can no longer properly imagine it.

There they found wolves and bears, as well as reindeer, moose, lynxes, beavers and wild boar. They also found snow and long, hard winters. Hunter-gatherer survival was tough in such harsh conditions and, difficult to grasp though it may be to us now (especially those of us who dream of recreating such wilderness), they may well have thought little of this, for it seems that they scuttled back to the coast, to the relative comfort of their fisher-gatherer settlements where they could build tidal fish-traps, pick oysters and mussels from the shoreline all year round, and where hunting was the simple matter of exploiting the abundance of unwary ducks and gulls. They also found seals – grey (or Atlantic) seals – tens of thousands of them.

The seal was a godsend to Stone Age man; the answer to so many of his prayers – if, that is, he was given to some sort of prayer, which, judging how we felt about putting to sea that morning, seems highly likely. The seal was everything you could ask for, if your perceptions and ambitions were circumscribed by Stone Age man's perpetual struggle for food. Here, in one relatively easy-to-catch cigar-shaped bundle, was a tough waterproof hide, excellent for clothes and coracles; blubber for keeping out the cold and for rendering to valuable oil; a large quantity of high-protein meat; sinew for thread; useful bones and, most bountiful of all, every year lying helpless along the tide lines of many beaches – there for the taking – was a glut of all these delights wrapped in the soft, fleecy fur of newborn pups. Not so surprising, then, that what we know of Stone Age man in Scotland comes from his campsites, caches of tools, bones and shell middens, all located on the coast.

It is a plausible hypothesis that the grey seal assisted early man in his settlement of Highland Scotland, principally throughout the islands. Seals could have been just that extra advantage nature sometimes provides to clinch the successful colonisation of a niche – good for Stone Age man, less so for the seals. They faced a problem. Their biology required them to come ashore to mate and to give birth to helpless pups on the beach. For a month every year the pups were immobile, defenceless and unable to escape to sea – so much easy meat. To make matters worse they were programmed to rut in the autumn and give birth in the spring – in good weather, long daylight and calm seas – all facilitating man's predatory access. To these early men the seal harvest was, quite simply, a bonanza.

But grey seals, *Halichoerus grypus* (the Roman-nosed sea-pig), are intelligent mammals. After several thousand years of Arcadian isolation in these islands the seals now found themselves severely disadvantaged by this unexpected turn of events; they had no alternative but to abandon their favourite beaches and haul-outs for remoter, safer islands, of which there are many off the west coast of the Highlands. Predictably, man followed. It can be argued that by this process and aided by the continuous supply of seal products, Mesolithic man dispersed himself throughout the Small Isles and the Hebrides, perhaps Orkney and Shetland too. As the seals moved further and further out, so did the hunters. With their vessels they could continue to exploit the skins, meat and bones, and look forward to a seasonal harvest at pupping time. And, as is man's interminable wont, our Stone Age predecessors overdid it. Had marine biologists and conservationists been around at the time, they would probably have forecast the ultimate extermination of the grey seal in Scotland.

But the seal's innate intelligence is not merely perceptual; it is also biological. Deep within its genes lay a trump card known as 'delayed implantation' (actually not uncommon in mammal biology). This is a physiological process by which the brain and the body come together to say, 'We're in deep trouble here and if we are to survive, we must do something about it.' After mating the normal process is for the embryo pup to implant in the uterus wall and steadily develop throughout the nine months of grey-seal gestation until birth. For most of the grey seals in the world that is what happened then, and still does today – but not for the Scottish and Hebridean populations. They have done something else.

Newborn grey-seal pups can't survive for long in water. There is nothing the adults can do about that in the short term, although it is interesting to note that the common seal, *Phoca vitulina*, has evolved to give birth in the sea, in sheltered water, the pup never needing to come ashore. This is a significant evolutionary step forward that will have taken thousands of generations to perfect. Oppressed by Stone Age man all those years ago, the grey seals chose a different and rather more drastic adaptation that may have provided a faster evolutionary response to their crisis. What they could do within a few generations – what they *did* achieve in the Scottish population – was to delay the implantation of the embryo so that the pup was *not* born in the late-May/early-June Highland spring, vulnerable and accessible, but at precisely that moment in the year when man was least likely to be able to get to the outlying islands. Simultaneously, they sensibly migrated their breeding grounds to the most far-flung, storm-battered, uninhabited islands they could find, where many of them have remained to this day.

The real success of this device seems to have been the timing of the pupping. The autumnal equinox falls on 23 September. At this moment in the year there is a reliable pattern of winds and spring tides emerging from the currents and prevailing winds in lower latitudes, which build up and lash their way round the northern hemisphere. In the Hebrides these sudden and often fierce squalls make seafaring particularly hazardous – in fact, the very weather in which you *don't* want to be putting to sea in your coracle or dugout canoe. The synchronism of the equinoctial gales and the new season of grey-seal pupping, late September to early October, was arrived at by the continuous process of trial and heart-rending failure – by natural selection, the only way that nature knows how to work. The hunting pressure must have been so great and so constant that the wretched seals were forced to try every option available to them. Only those that succeeded in delaying implantation would survive to breed; all the rest perished. Those that failed to move to pup on outlying islands were either killed or were perpetually prevented from breeding.

Just how long it took for the selected solution to establish we cannot know. What we do know is that Hebridean grey seals still pup in the autumn – a delay of five months – and then, immediately after giving birth the cows go straight into rutting harems to mate. They have opted to pit their survival chances against the angry seas, rather than face total extermination at the relentless hand of man, while birth for other populations (in Pembrokeshire, for instance, where fertile soils were far more important to early man than seals), remained in the spring, unchanged, with a separate haul-out to mate in the autumn. To this day, long after the hunting pressure has disappeared, you still have to run the gauntlet of Hebridean storms and high seas to see grey-seal pups in their far-flung colonies. So that is where we were going.

This little expedition was work. I was taking with me five young naturalists in their twenties, field-centre colleagues who were keen to witness this rare natural phenomenon for the first time. It was part study – an opportunity for the young staff to broaden their knowledge – and part team-building exercise. It would be a bracing challenge; an experience designed to catapult them out of the unrealistically cosy world of fair-weather natural history – that of butterflies and newts, of soaring golden eagles and the heraldic osprey snatching trout from Highland lochs – the wildlife that shaped their everyday lives. This would be different: wild nature in its bleak, unforgiving element, pitting and testing. It would demand endurance in the teeth of real physical discomfort, even the possibility of danger from the sea and the seals themselves.

When Hermione overheard me planning the trip, caught us poring over charts and listened wide-eyed to my warnings, she asked if she could come with us. At first I didn't take her seriously, dismissing her on the grounds that I couldn't mix work with family fun and that anyway, it would be far too rigorous for her. To my surprise she was undaunted, offering to run errands, even help with the washing up, if only I would let her join the crew. I denied her again and again, but she nagged and begged and looked so forlorn that at last I weakened, sternly warning her as I did so that the chances of a rough ride were *very* high, that seasickness was *not* fun, and that far from being the cuddly animals they are so often depicted to be, seals are potentially very dangerous.

I thought that if I made it sound grim she would falter and withdraw. I was wrong. She never flinched; the notion of risk seemed only to strengthen her resolve. Once I had given in I realised that the timing was probably right – it was time to expand her horizons too. However testing it might prove to be, it would be an adventure she would remember for the rest of her life, and a physical and psychological challenge that at nine years old I reckoned she was probably just about equal to. Besides, we had chartered *Pippin* because she was an excellent sea boat and because I had absolute confidence in Bobby, her friendly skipper, who knew those capricious island waters nearly as well as the seals themselves. Hermione was coming with us, but I was loath to tell her mother just how uncomfortable it might be.

In a chain we passed our kitbags down the rusty quayside ladder and pushed off. The engines churned and *Pippin*'s high bows nudged into the wind. While the team stowed their kit below decks and chose their cabins and quarters, Bobby and I picked over the charts and discussed our passage up the Sound of Mull to our first night's sheltered anchorage in Loch Aline. It would take us a whole day's cruise to reach Treshnish, taking the north route, up the Sound of Mull, past the picturesque natural haven of Tobermory and out, round Ardmore Point, into the eye of the wind. With the long, low islands of Coll and Tiree on our starboard bow we would bash our way down the north-west coast of Mull to Lunga, the main island of the Treshnish archipelago. Here, in a dangerously inadequate anchorage – just a few square yards of slack water – there was only the low bulk of the 337-foot Cruachan Hill to shelter us from 3500 miles of the North Atlantic; open ocean all the way to the Labrador Basin. This was exactly what the seals had planned. No one knew better than they just how quickly and savagely the west wind could whip the ocean into gale force 8 or storm 9 or 10. None knew better how those seas would lash and thrash at the sea-shattered basalt rocks where, almost perversely, they had systematically chosen to give birth to their pups.

Tucked in behind the lumpy mountains of Mull our journey to Loch Aline was smooth, the sea glossy and black. In the falling darkness we crept into our anchorage as a soft rain sizzled all round us. We ate supper sitting round the galley table; the tide caressed the hull like a lullaby and the shrill piping of oystercatchers echoed from the shore. Whisky glasses chinked and beer cans coughed and spat. The portholes fogged up and laughter rang out. Hermione had stayed on deck throughout the journey, seeming to love every lurch and roll of the gentle waves; her cheeks flushing lobster-red in the sharp salt air. 'Do you want to turn for home?' I called out teasingly to the rain-bedraggled figure in the bows.

'Don't be silly, Daddy,' was the rebuke I deserved, my daughter not even bothering to glance back at me.

She was asleep on the padded saloon bench before supper was over. I eased her gently into her sleeping bag and left her where she lay. Before turning in I walked the deck in the still night air. A gibbous moon slid between clouds, its old silver-gilt shimmying across the surface of the loch. The west Highlands is narcotic, a drug I am drawn back to over and over again. Its gentle setting of lochs and mountains exalts the soul, while the sighing wind and the fluting calls of eider duck rising and falling from far out on the water can be as elegiac and poignant as a weeping violin. For the first time that evening I glimpsed the possibility of many such expeditions into the wild with Hermione. If I could somehow pull this one off, I knew there would be no stopping her. As I ducked through the hatch to go down to my cabin I felt a warm glow of satisfaction. Our first little adventure together was under way.

At dawn the stern voice of the shipping forecast accompanied us north-west up the Sound of Mull into a brisk wind that continued to freshen all morning. To Hermione's delight the sea began to break over the bows. In scarlet oilskins over green wellies, with a gaudy orange lifejacket strapped on top, she stood braced in the bow cradle facing into the waves like a figurehead, daring them to douse her. *Pippin* plunged through the undulations that hissed and seethed, foaming and angry at each curling crest. Every few minutes one would smash into the bows as they dipped through the troughs and a wall of white water would surge over Hermione until, with numb fingers and a face almost as bright as her oilskins, she gave up and staggered below deck to dry off.

When at last we rounded Ardmore Point and turned west to run down the Mornish coast of Mull towards the Treshnish Islands we met the full force of the Atlantic swell urged along by a stiff westerly breeze. Now the brave little *Pippin* pitched and heaved on every wave and trough. The hull slammed down into the water and the bows crashed into the oncoming wall like an icebreaker. Below decks pots and pans had to be clamped in place, and glasses and crockery leapt from their racks and smashed on the floor. I took the helm while Bobby frantically stowed and secured everything he could. After an hour of this battering I wondered whether he was beginning to regret the whole undertaking. The rest of the team sat grimly in the deck saloon, clutching on to each other and their seats as best they could. As I grinned at Hermione I saw that a green pallor had diffused into her cheeks and her face was now set in a smile of gritted teeth.

As the morning wore on the sea eased back to a far more comfortable chop. The tide turned in our favour and our speed visibly quickened. Lunga and the Dutchman's Cap, the two most prominent Treshnish Isles, stood out as dull smudges above the bows. Soon we could see the distant white rim of surf clawing at their cliffs. Our spirits rose and the radio crackled good news. Barometric pressure was rising; winds began to fall away. The sun came struggling back, elbowing its way through towering Himalayas of cumulus cloud with leaden underbellies and shining heads. The world turned white and blue. For the moment at least, Poseidon was smiling upon us.

We ran into Lunga on the north approach. In sheltered water the sea was suddenly unexpectedly calm. The tide was low and we had to throttle right back and watch the echo sounder with an eagle eye. Jungles of kelp warned of shallows on either side of the channel between outlying reefs and islets. As we crept in, black guillemots scurried away from us like tiny red-legged penguins. To our surprise and delight, from one of these low, grassy-topped outliers a great flock of barnacle geese rose up in alarm. The sun streamed from their black and silver backs and the air vibrated with their high-pitched yelps. There were more than three hundred of these majestic wild geese just in from their Greenland breeding grounds, en route for the rich grazing of Islay and other fertile Hebridean islands for the long winter. They had stopped to feed and rest in the haven of these uninhabited islands before pushing further south. They were unexpected and thrilling; their music lifted us with them into the blue dome as they circled round in a broad ring, pitching once more as we slid gently past.

Inside the islands the surface of the sea shone like rippling silk of brilliant lapis lazuli. Cormorants slapped away from us, leaving trails of pearls glowing in their wake. Herons rose from the rocky shore barking their displeasure like regular bargees as they hauled away to quieter coves. We manned the bows on both sides, nervously eyeing the rock-studded channel; yard by yard Bobby's voice sang out the echo soundings. The engines purred restraint. Now entirely recovered, Hermione sat cross-legged on the saloon roof, well out of the way, binoculars in hand, eyes dancing with excitement. Later she told me she felt like Jim arriving at Treasure Island aboard the *Hispaniola*.

At Lunga we dropped anchor. The engines died. The sun was bright and the wind had all but vanished. We were standing off a hundred yards from the hump of Cruachan Hill and about half as far again from a long, low island braced against the Atlantic, accurately named Sgeir nan Chaisteil – the castellated rock. The island is basalt, an igneous rock solidified from volcanic eruptions in the earth's crust. The earth's liquid core oozed into the light and set rigid in dark, crystalline columns, as in the famous Fingal's Cave on Staffa, which was a distantly visible smudge a few miles to the east. Sgeir nan Chaisteil's basalt is deeply fissured, shattered like the ruin of an ancient fortress by the ravages of wind and waves. Its outline, crenellated into battlements, has handed the island its Gaelic name – a final, craggy outpost of land and language.

As the engines died away we became aware of a strange, mournful wailing filling the bay. A wild concert of bassoon and cello rose and fell around our ears, mostly echoing across the water from the wave-cut platform facing us on Sgeir nan Chaisteil. This was seal song, once heard, never forgotten. Like the stag's roar, it claims the listener and the land so that you can never view that landscape again without it bursting in upon your consciousness, dragging you back to that first experience. For me it is a sound that belongs exclusively to wild Hebridean beaches. It embodies the elemental qualities of the wind and the waves in its fluting melancholy, as haunting as a grief, bearing with it all the accumulated history of a thousand generations of predation at the hand of man.

The land was littered with the fat cigar shapes of seals hauled out on the wave-cut platform and the beaches. With binoculars we could see them on the rocks on both islands, way above the tide line – some of them lazed like corpulent sunbathers on green grass. Glossy blue-grey heads with long Roman noses bobbed in the shallows and three or four inquisitive individuals popped up beside us to inspect our boat. These seals trod water a few yards off; we could see the dappled patterns of their sleek, wet fur beneath the water as their W-shaped nostrils nervously flared and clamped. They eyed us with suspicion, wide round orbs staring out at us, hovering on the edge of fear. Every now and again one would panic and dive suddenly and clumsily, with a snort and a great swirl of green water, taking the others with it in turmoil. For a few seconds we could follow their progress; ghostly and serene they swept beneath the hull in an ethereal, curving ballet, rolling and spinning with effortless grace. It was as though they had become fluid themselves, assuming the lambent properties of the rippling tide rather than those of a huge mammal – and then there they were again, a shining head in a ring of sea, in a new position, silently eyeing us all over again.

Hermione had seen seals many times before. Grey and common seals are present on the mud and sand banks of the Beauly Firth only a few miles from home. But never this close; never before had she witnessed this consummate grace and their sublime adaptation for the deeps. She lay flat on the deck on her tummy, her face over the gunnel, only a few feet from their wide eyes and inquisitive faces and their sleek, shadowy forms as they curved below us. 'Aren't they amazing?' I asked her quietly as I sat down beside her. She nodded emphatically, saying nothing. I knew then that it had not been a mistake to bring her.

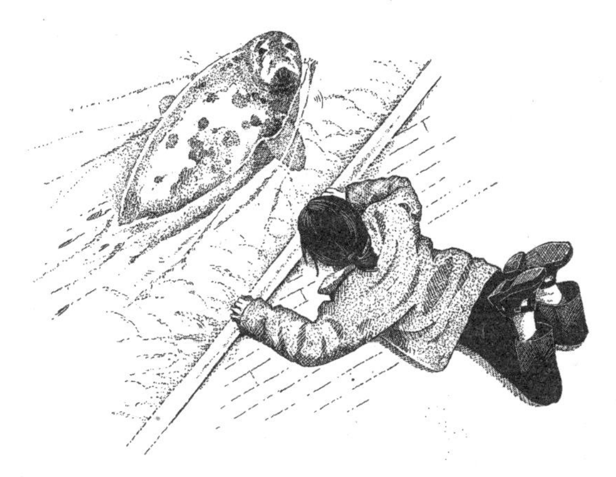

We lost no time in getting ashore. We held another short briefing on how dangerous grey seals can be, and Bobby warned us with an uncharacteristically stern face that he was extremely suspicious of this lull in the weather. If the wind got up he would have to depart fast. Lunga is no anchorage for a gale. If we didn't return when the ship's hooter sounded he would be forced to leave without us, marooning us for the duration of the storm, perhaps for days. The natural harbour in the tiny island of Gometra is the nearest safety, fully an hour and a half away, even with the wind at our back. We had to keep the *Pippin* in earshot at all times. The rubber dinghies were lowered and we left our skipper in his wheelhouse with an ear glued to the radio. As we rowed the short distance to the shore I suddenly had a curious sense of misgiving. It wasn't hard to imagine that these sunlit, calm conditions might change quickly.

I had been to Treshnish several times before, so I knew the best pupping beaches and where the huge bulls haul out to claim their rutting territories. We decided to go to Sgeir nan Chaisteil first because the wave-cut platform in the lee of the escarpment is the best site. The pregnant cows choose this for easy access: smooth rock to clamber over, which only floods at very high tides. This means the pups can be born above normal high tide, protected from the prevailing winds, and can be tucked in behind boulders or in one of the many crevices that bisect this basalt shelf. The cows can come and go at will, shuffling over the smooth rock with ease. It was perfect for us too. From the battlements we could look down on the cows and their pups without disturbing them. It was a natural viewing gallery; I can think of nowhere better designed to witness this rare congregation of natural history in the raw.

We shuffled into our positions on our bellies. Hermione thought this was huge fun and she was better at it than most of us, slithering like an eel across the rich quilt of grasses and wildflowers of this unexpectedly fertile island soil. Lying there side by side in a row we seemed to be emulating the seals below us. My daughter could scarcely contain her excitement. She was raised up on her elbows as she scrutinised them through her binoculars; the delicate cloud and sea-mottled patterns of the dry fur of the cows and the fleecy white of the newborn pups at their sides. 'How about that?' I whispered, trying to conceal the smugness of a promise fulfilled. She didn't respond immediately. She was too busy. I had to learn to hold back, to resist claiming the glory for myself. I could see that she was transfixed, drinking in this extraordinary scene in extravagant gulps of new experience. Then she asked, 'How close can we go to the pups?'

'Very close,' I answered. 'But later on, when we go down to the beach, we can go right up to them.' The smile she threw me and her flashing eyes said it all.

The grey seal is Britain's largest mammal. It is huge; weight less important in a cold marine environment than the insulation its thick blubber provides. Mature bulls are commonly 630 pounds and have been recorded up to 900 pounds, a gargantuan by any standards. Cows are much sleeker at only 450 pounds, but still more than twice the weight of a large man. The pups are born about the size of a human baby, thin and wedge-shaped for a rapid birth, head first, one of the fastest births in nature, between ten and thirty seconds. Their white shaggy fleece affords them initial protection, but in just a few hours they have laid down so much blubber from their mother's rich milk – up to 35 per cent fat (3.5 per cent is the maximum fat allowed in cow's milk) – that they are barely recognisable. Without this layer of insulation they would quickly succumb to hypothermia in the wind and the rain, and especially the several hours of high tide each day when they are lashed by spray or washed off their beach by waves. When this happens the cows have to be there to prevent the pup from drowning: they have evolved a particular behaviour pattern of protection, literally blocking the waves from the pup with the length and weight of their bodies.

For half an hour we lay watching the comings and goings of the seals below us. Off to our left two bulls jostled and harangued each other for territory. The larger bull, parked firmly on the beach and evidently in possession, snarled and lunged at the incomer, who was partially immersed in gentle waves. They faced each other down; every few minutes the challenger came again, thrashing the water to foam as he surged ashore. He lunged at the bigger bull with bared, jagged teeth, lacerating the rolls of blubber around his neck and shoulders as he did so. His opponent lunged back as they thrashed about nose to nose. Blood flowed. Hermione winced, turning to look at me with horror, pointing in case I hadn't seen that the edge of the sea was gory with the blood of both combatants. 'Don't worry,' I reassured her, 'it doesn't matter to them. They always do it; it's like stags clashing antlers or rams charging each other. The salt will quickly heal their wounds.' This lunging and neck slashing was of no

consequence; superficial wounding that may well have added to the stature of the dominant bulls, who seem to wear their old scars like an Aborigine does his face paint.

One of our objectives was to estimate the approximate size of the Treshnish seal rookery, so we decided to split into pairs and head off in different directions. Hermione and I took one of the dinghies back to Lunga to tackle the long beach below Cruachan Hill where there is no wave-cut platform. Here, seals have to struggle steeply up over bladderwrack and shingle to a tide line littered with huge boulders, some the size of houses. It was here within this maze of crannies and chasms, that the Lunga cows had chosen to give birth. Hermione rose to this new challenge with excitement and relish. She had longed to see newborn pups close up and now was her moment.

Our progress became a game of grandmother's footsteps. As we rounded the edge of a large boulder, we scanned the ground ahead and proceeded as silently as we could so as not to alarm the cows. Sometimes we missed one; some were lying so still that we mistook them for the rocks. Occasionally this happened right beneath our feet so that we jumped with fright as a cow lurched suddenly forward, levering down the beach with powerful fore flippers, floundering through the shallows with a great commotion, abandoning her pup to its fate. Once in the sea she would turn and tread water, rising head and shoulders out of the sea, straining to see what we would do. We moved on quickly.

Hermione found this whole process thrilling. She stuck close by me and became very expert at spotting the sleeping cows ahead. When we found unaccompanied pups we tiptoed in to take a closer look, approaching to within touching distance. They lay on their sides with eyes tight shut, occasionally sighing audibly through spread nostrils. When they were awake, tears immediately formed in their eyes, to prevent the cornea from drying out. It made them look tragic, as if they had been abandoned. Hermione wanted to pet and cuddle them – an entirely predictable reaction to their endearing puppy faces and evident helplessness – but I had to persuade her that, quite apart from their sharp milk teeth, wildness is a precious quality to be respected and honoured. Just once I allowed her to extend a hand toward the fleece of a tiny pup, probably only a few hours old. She knelt quietly on the shingle at its tail flippers. Its wide, wet eyes watched as her small hand eased forward. Quite suddenly it turned to face the hand and hissed a throaty, aggressive cry, baring its tiny teeth as it did so. The hand sharply withdrew and a shocked face turned to me: 'Why's it doing that?'

'Because it's afraid of you and wants to defend itself.'

'But I won't hurt it.'

'No, but it doesn't know that and most humans seals have met have only been interested in killing them.'

'Oh, I see,' she said quietly, as we backed off to let its mother return. We have to discover our own boundaries and limitations in nature, to comprehend and interpret its subtle, often beguiling ways.

We clambered over a long vein of smooth bedrock that bisected the beach like a stranded whale, snout into the hill. From its top we sat and scanned the way ahead through binoculars. Cows were dotted about in dozens; about a hundred yards further on there was a little knot of them, perhaps as many as twenty strewn across the rocks like barrels, jettisoned cargo that had washed ashore. We decided not to disturb them and gave them a wide berth. To achieve this we had to head up the beached whale rock to the grassy headland where all the flotsam of high spring tides had arranged itself in long lines of rank detritus. Clumps of thrift had flowered here all summer, their delicate pink blossoms still nodding between black mounds of dried-out wrack and kelp, bleach bottles, polythene beer-can holders, green nylon netting and tangles of frayed orange rope.

Man's profligacy imposed itself upon us like a bad smell. The mess offended Hermione: 'Why do people do this?' she asked. 'Don't they care?'

'It's probably more that they just don't think.' I was pleased by her spontaneous responsiveness. After years of continually meeting omnipresent pollution in wild places I expected it, and no longer commented. It was good to hear my daughter speak out, I felt proud and a little ashamed that I had said nothing myself. The seals were far below us now and we relaxed, ambling along hand in hand, chatting and beach-combing, turning a fish box here and a collecting a crab claw there. We were no longer paying attention.

We had climbed over the first of two large wedge-shaped rocks, massive slices of basalt that had fallen on to the beach. They were both some seven or eight feet high at their seaward end, and many yards long, tapering right back to the cliff from which they had split. Between them lay a deep chasm of shingle the width of a country road. I slid from the rock on to this narrow little beach and turned to help Hermione down. She was transfixed; frozen to the rock, pointing and grimacing with something close to terror in her eyes. There, only a few yards higher up the narrow beach from where I stood, lay six vast, sleeping bull seals, snoring like drunks. They were trapped between the sheer walls of the rocks, a private secluded beach away from all the others.

For whatever reason, these bulls were not in rutting condition (their turn was yet to come, different animals probably come into the rut at different times, just as the cows spread their date of giving birth over two or three weeks to ensure that they are all re-impregnated), so they had shuffled up to this secluded chasm, well away from the cows and other bulls commanding territory on the shore, to lie out together in a smelly, gut-rumbling and farting huddle, almost touching each other, and, mercifully, fast asleep. We were very close. I was blocking their only escape route to the sea. Instinctively I knew that I couldn't climb back on to the rock without waking them. I would have to retreat on tiptoe, down the strip of beach, round the high shoulder of the rock, and back the way we came. For safety Hermione had to stay on the rock. I whispered to her to move to the highest end and sit very still. As soon as she was safely in place I began to creep away.

Whether it was the soft crunch of stones beneath my feet or our wafting scent that woke him, I shall never know. I heard a frightened 'Daddy!' from above and turned to see the nearest bull sitting bolt upright on his fore flippers, nostrils flaring, double chins trembling, eyes white-rimmed with alarm. With a snarl and a lunge of power that sprayed shingle in all directions, he launched himself towards me like a landslide. All five others were instantly awake. The earth seemed to be heaving; something volcanic was tearing the beach apart. They came like a lava flow, hot and unstoppable, outraged, roaring, snapping slavering jaws, a plunging wall of blubber and vicious, bared teeth surging at me faster than I could run. I had a start of six, perhaps seven yards. I sprinted to the high wedge of Hermione's rock and made it round the end, out of their path, just in time. They stampeded past. With my heart pounding in my chest I reached the other side to look back and see a seventh huge bull, so far invisible because he was tucked under the thin end of the rock wedge, mount the rock like a tidal wave and lurch towards Hermione.

There was nothing I could do. She was out of my reach and I could only stand and watch this colossal beast weighing well over half a ton charge straight at my nine-year-old daughter. I yelled, 'Jump!' but it was pointless. Rigid with fear, she was a little red blip stuck to the rock, as vulnerable as a sea anemone. The bull surged on. It brushed past her at a few inches' range, and plunged over the rock's end, falling seven feet to the beach with a sickening crunch on to the shingle, which sprayed wide from the impact. It somersaulted in an ungainly mass, recovered, and frantically heaved its way on down to the sea, where it disappeared into the turbulence, still foaming from the others. The seas had parted as they ploughed in, as if seven ships had all been launched together.

We hugged until her tears and our combined heartbeats subsided. All she could say was, 'He had bad breath!' I laughed. It took her a little longer to see the joke, but I was laughing to conceal an overwhelming desire to weep. You may imagine that I have dramatised this incident for effect – but not so. I have recounted it just as it happened, one of those unaccountable mishaps that will always punctuate any active and adventurous life. It goes without saying that I would never intentionally place my own or anyone else's child (or myself, for that matter) in a situation of self-evident harm. But I believe that experiences such as this – provided, of course, that they end happily – serve a valuable function in establishing limits and understanding of wildness and wildlife. Just as in my canoe I had discovered the power of swans, so Hermione had vividly experienced the terrifying might of grey seals when, in our case inadvertently, you place yourself between them and the sea.

When a child at play jumps off a bank, falls and grazes his or her knee, limits are subconsciously logged about banks and safe jumping distances. To prevent grazed knees a parent would have to ban all banks and all jumping – patently an absurd restriction. The same applies to any sport or other activity; the whole purpose of sailing, rock-climbing, mountaineering, riding horses or any of the dozens of other outdoor pursuits, is to broaden horizons, build confidence and skills, to learn to handle emotions such as fear and panic, as well as to deliver up the spiritual and intellectual benefits of real and challenging experiences.

The first-hand experience of wildlife is much the same. Mankind has gone to extraordinary lengths over many centuries to distance himself from the wild, including trying (pretty successfully) to exterminate animals seen to be threatening or dangerous. Wolves have been demonised in Western culture, despite there being virtually no hard evidence that they kill human beings. So effective has this purification process been that the very thought of wild animals prowling round our houses at night is now utterly alien to most people – a perceived threat where in reality none exists at all.

I believe that reversing this cultural direction, actually seeking out wildlife and directly observing it – learning to revalue it – is vital if we are to retain any breadth of biodiversity on the planet. Just as a musician must hope that his children will grow up understanding music, I have always wanted mine to know and understand wildlife, to conquer irrational fears and replace them with comprehension, with real knowledge. I have always viewed it as fundamental environmental education – a discipline I believe to be essential for human survival – extended to a personal and subjective process of familiarisation. Would I still be arguing this point if my daughter had been badly injured or even killed by that huge seal plunging past her? I hope I would. Life is for living.

Yet there is a thread of reality that is missing from this tale. It goes back to the Stone Age men and hundreds of generations of their Hebridean descendants who have come here throughout history, not to observe and count seals, but to kill them. I doubt that many human beings have actually died in the jaws of grey seals over the past few thousand years. A hunter would have to corner the animals and be physically attacking them to elicit such an aggressive response. No doubt a few people did get badly bitten in the mêlée of the bludgeoning gang attacks that must have been the norm – and that would be a most painful experience, even possibly resulting in fatal infection. But, of course, the numbers of seals killed by these raiding parties and the many more wounded with clubs, spears and harpoons that will have escaped to sea and survived to pass on their dread of mankind, is enormous – simply countless.

Seals have learned over the millennia that man is to be avoided. Those huge bulls, terrifying though they seemed to us that day, had no thought of attacking Hermione or me. Had I stood my ground I might well have been knocked aside, very likely injured by their surging, gigantic bulk, but they would not have lingered to savage me. What they so desperately sought was escape to the ocean. It is there that they have invested their genetic energy in hydrodynamic adaptations; and it is only there, in the welcome cool of its green deeps, that they really feel safe.

No sooner had we recovered from this drama than we heard *Pippin*'s foghorn booming across the bay. We could see why. Waves were slapping and creaming at her bows as she bucked and plunged at anchor; her ensign flickered in a rising breeze, backing toward the south. We knew very well what this meant. We scurried back to the beach where the others were assembling at the dinghies. Launching was not the same as landing. Now there were angry waves to push through, waves that caused the dinghy to leap and lurch. Spray drenched us and filled the boat; water sloshed around our feet.

I took up the oars and struck out for *Pippin*, but it was not to be so simple. The tide flooding between the islands and a tugging crosswind turned us sharply away so that fighting against it, I quickly tired. Every stroke swept us further along the shore, away from *Pippin*. In just a few minutes I was exhausted. There was nothing for it but to return to the beach, disembark, drag the dinghy a quarter of a mile upwind and launch again so that we could run down on a diagonal to the ship. By the time we reached her, the anchor was up and the engines held her bows into the current. Bobby was palpably relieved to be under way.

Ten minutes out we left the shelter of Lunga. The wind slammed into us and we hit a sea like something from *Moby Dick*. From the mid-Atlantic huge ocean rollers bore down upon us, row upon row, fifteen feet from dark trough to breaking crest, so that the little ship disappeared into deep valleys of foaming sea. There was no question of standing in the bow cradle now. The bows were barely visible as water surged down the gunnels. Bobby's face looked grave as he wrestled with the helm; the rest of us clung on whilst we ran before the howling wind, as directly as the valiant, pitching little ship could manage, for the island shelter of Gometra.

The speed with which the gale had appeared was frightening. When we rounded the headland into Gometra's natural harbour we heaved a huge sigh of relief. We found an anchorage as close to the shore as we dared and laid out all the chain we had. Bobby refused to close down the engines for fear of dragging the anchors in that blasting, purging wind. The waves had gone, but the wind was relentless. The surface of the bay chopped and flurried, gusts blasting across it, whipping the dark sea into whorls of turbulence. Throughout an uneasy supper the rigging whined and clanked incessantly, like the unyielding looms of a textile mill. As darkness hemmed us in I knew we were in for a sleepless night. Bobby never made it to his bunk. In the event the anchors did drag and we had to re-lay them twice, nudging ever closer into shore and praying that the wind wouldn't veer and push us on to the rocks.

Only Hermione knew nothing of it. Exhilarated by the roaring storm she had stayed on deck as long as she could, no suggestion of seasickness now, until she flopped into the saloon, exhausted from perpetually clinging on but still smiling ecstatically. 'What a day!' I commented as I helped her unravel from her dripping oilskins.

'Brilliant!' was all she could muster. She had crawled into her bunk, almost too tired to stand. When, a little later, I climbed down to say goodnight, it was too late – she had gone. In her dreams she was a seal pup, somewhere out there in the lashing darkness, battling for its tiny life against the unforgiving tide and the pounding waves.

Daybreak hailed the journey back through the same crashing seas, past the slanting, columnar grandeur of Fingal's Cave on Staffa, where it is said that in the time it took for his cruise ship to glide by, and from the waves booming into the cave, Felix Mendelssohn extracted the essence of his *Hebrides* overture. We saw sunlit gannets wheeling and nose-diving like kamikaze pilots, and we returned to Lunga to hear once more the mournful song of the seals. Yet again we were forced to run for cover as another gale whirled in from the ocean. It was as though the seals had not properly made their point, as though there was now a ring of mockery in their song. Perhaps we had overlooked that once their creative biology had combined forces with the wind and the waves, while few men may have died in the jaws of seals, many must have drowned in the ceaseless struggle to defy nature's enduring solution to their greed.

SEVEN

A Love of Life

> . . . the multifarious forms of life envelop our planet and, over aeons, gradually but profoundly change its surface. In a sense, life and Earth become a unity, each working changes on the other.
>
> LYNN MARGULIS, 1938–

The ocean rolls ashore in slow thunder. On the great shingle beaches of Dorset and South Devon it's a thunder underscored by the snarling of pebbles being rolled to infinity. The ocean seems to be breathing stertorously at its very edge. After each roaring exhalation both shingle and waves are sucked back through bared teeth.

The great enigma of Chesil Beach faces the wide waters of Lyme Bay. It sweeps between the two lighthouses of Portland Bill in the east and Start Point, the most southerly tip of Devon, a dim blur sixty miles to the west. In this heaving amphitheatre a savoury wind purges the ego like holy water. You stand poised at the rim of a vast font into which you feel God is about to plunge you, whether you like it or not. An arc of creaming breakers is all that separates you from the sensuous, undulating vastness that presses ever forward.

Far away, where the horizon ought to be, an indistinguishable grey sky conspires with the waves and comes at you again, but overhead, so that you feel as though you are about to be smothered. Somehow the sea has split itself evenly, donating half its depths to the clouds. Beneath your feet, not only is the expanse of banked shingle a rare and remarkable phenomenon in itself – *terra* not the least bit *firma* and endless fun for a child to crunch noisily into – but it also provides an exhilarating platform from which to be swallowed up by this awesome seascape.

No one has quite fathomed the process by which the waves have created Chesil's twenty-mile beach. The precise grading of the flint pebbles is intriguing, especially so to a total stranger like me. It rises to forty-five feet above high-water mark and extends some eight fathoms below low-water springs: a total of nearly a hundred feet of banked pebbles, two hundred yards wide, edging the land in a steep, shifting crescent from Bridport to Portland Bill. It's unique; nothing quite like it exists anywhere in the world, insists Professor J.A. Steers of Cambridge University in his classic work *The Coastline of England and Wales*. What is so unfathomable is the arcane and pedantic grading of the pebbles. Above sea level they increase in size from north-west to south-east; below the waves they mysteriously reverse. At the west end, at Bridport, the stones are like hazel nuts; at Portland they have swelled to become swan's eggs. To walk east from Bridport is to start out in gravels that shift with every step; by the time the white finger of the Portland Bill Light looms out of the salt haze you are stumbling over heaped cobbles. If it were possible to do the same thing underwater a few yards offshore, precisely the reverse would be the case.

In April 1964 a gang of enterprising Cambridge students of physical geography dumped a lorryload of random-sized bricks between the tide lines to see if the waves could be duped – a purposeful academic prank, a geographer's April Fool performed with all the naïve optimism of King Canute. The waves weren't impressed, far less fazed. In swimsuits and masks the students stayed to watch what happened. The high spring tide came roaring in, bowled the students over and smothered them so effectively that they had to be rescued; then it withdrew with a chuckle. The bricks had gone. When eventually they found some of them again, they had been graded, rounded off and summarily dispersed to appropriate sites on the beach. No one has volunteered for such humiliation since, nor have they worked out quite what quirk of physics brings this autocratic natural phenomenon to bear.

To Hermione to sit on the bank's crest and invite gravity to pull you down through the steeply terraced shingle is irresistible. I join in. Every pebble is so clean, so dry and perfect in its naturally manufactured smoothness that you scoop them up in handfuls, letting them slip through your fingers like marbles. They invite you to lie back and wriggle your form into their cool, yielding bed, as friendly as a beanbag. Slipping, sinking and laughing we clamber back to the top of the bank and breathlessly fling ourselves down. We gaze out to sea. The salt wind keens our cheeks. Breakers crash and roar at our feet with indefatigable violence. The timeless ocean rages at us, at this insubstantial rim of land. But we're not here just for fun. We've come to Dorset on this blustery February day because a textbook has told Hermione that this is one of the best places in Britain to find fossils.

Along this coast, riled by winter storms, the ocean regularly takes great bites at the chalk and lias cliffs, freeing as it does so the petrified shells of countless sea creatures that have lain locked in their sedimentary beds since they perished on the tropical ocean floor 180 million years ago. We have been directed to Charmouth, a few miles west of Chesil Beach. This charming but otherwise unexceptional little village has discovered a market advantage in its fragile cliffs. It vies with neighbouring Lyme Regis as the fossil capital of the south coast. It sports a visitors' centre and museum where the resident geologist, in his fisherman's herringbone polo neck, encourages interest in palaeontology and helps you identify your finds.

Charmouth and Lyme have at least four fossil shops on their main streets, in the window displays of which vast, coiled extinct molluscs called ammonites appear to flirt with spread-eagled dinosaur skeletons cast in stone. Here you can buy almost anything geological from anywhere in the world: baskets of brightly coloured stones with evocative names like dolomite, chalcedony, gypsum, turquoise, olivine, topaz, tourmaline, agate and obsidian, clutter the counters.

Displays locked behind glass reveal forty-million-year-old ants trapped in amber resin, gleaming shark's teeth the size of arrow-heads, and Moroccan geodes cracked open to reveal crystal caverns of pink and mauve quartz. Polished whorls of prehistoric organisms have been sectioned to expose the intricacy of their inner designs as expensive and arresting ornaments. You can come to Charmouth and Lyme Regis and go away laden with fossils, without ever setting foot on the beach. But for the moment that is of no interest to Hermione. She has caught a bug far more infectious than shopping.

In the Highlands of Scotland, near her home, on the beaches of the Black Isle, she happened across a story even more celebrated in geology than the dramatic chalk landslips of Lyme Regis. Picking stones and shells like all children will, she discovered the coiled inscriptions of tiny ammonites, white and glyphic, on a fragment of old red sandstone. She might as well have found a gold doubloon. She was hooked.

A mile along the coast from the elegant Georgian town of Cromarty, it was at Eathie in 1830 that the young stonemason Hugh Miller first discovered a fossilised fish. This was entirely new to science. It would not only make Miller famous, turning him into a writer of note and a founding father of geology, but it would also challenge his deeply held religious belief of creation. A year later he discovered *Pterichthys*, the winged fish that was to be named *milleri* after him.

> It opened with a single blow of the hammer; and there, on a ground of light-coloured limestone, lay the effigy of a creature apparently fashioned out of jet, with a body covered with plates, two powerful-looking arms articulated at the shoulders, a head as entirely lost in the trunk as that of the ray or the sun-fish, and a long angular tail.

One *Pterichthys* he found had 'a strong hexagonal plate, fitted upon it like a cap or helmet'. The fish normally carried these dermal plates folded across its back, but in some desperate final drama they had been spread in defence. Miller wrote: 'We read in stone a singularly preserved story of the strong instinctive love of life . . .' – and '. . . a wonderful record of violent death falling at once, not on a few individuals, but on whole tribes.'

It is this 'strong, instinctive love of life' that has somehow inveigled me seven hundred miles from home on to this blustery Charmouth beach on a February morning. At the car park, where the turbid River Char bisects the beach and vanishes into the waves, you are faced with two choices: left or right. To the left, over the footbridge, stand Stonebarrow and the conspicuous Golden Cap, the cliffs for which Charmouth is best known; to the right is a smaller and much less attractive cliff ominously named Black Venn. We choose left. A handful of intrepid enthusiasts with haversacks, hammers, picks and shovels are heading off along the shingle beach as if they know something we don't. One young man even carries a crowbar. Hermione is on to it in a flash. 'Come on, Dad.' Eagerly clutching her yellow-shafted geologist's hammer, she tugs at my hand as though if I don't hurry I shall miss someone very famous who is passing through.

By-laws do not protect the fossils here as they do in some other sites. There's no point in banning collectors from these cliffs. Winter storms and high tides thrash the friable cliffs of blue lias, grey limestone and greensand, so that they constantly crumble and slide on to the beach. Here they are washed out by the waves and quickly dispersed, graded by size and density, like Chesil Beach, to be ground, rounded and ultimately eroded to sand. To insist that fossils naturally subjected to this destruction should not be collected is patently ridiculous, so the enthusiasts come in droves – a fossil-rush. Some even do it for a living, supplying the fossil shops, paid by the ammonite foot and the micraster inch.

There has been a storm. The car park is strewn with shingle and driftwood, and a huge chunk of concrete revetment has broken away where the beach has washed out beneath. Ragged plastic bunting has been strung around it to warn us off. It flutters half-heartedly in the salt breeze with precisely the opposite effect. Children are immediately drawn to the edge so that panicky mothers rush after them and drag them away. Herring gulls sit about dejectedly, not a picnic in sight. The litter bin has blown over and a metal ice-cream sign bolted to the concrete is wind-bent.

It is the storm that has brought out the fossil hunters. The heavy rains of a frostless winter have caused mudslides from the cliffs, which drip and ooze like wounds. It is to these that the hunters hurry, to harvest the fresh crop of fossils exposed to daylight for the first time since they sank into the Mesozoic mud all those millions of years ago.

A mile along the beach we arrive at a huge chute of blue-grey marl and muddy clay that has surged many yards out on to the beach from 150 feet above us. It's still coming, slowly and cloyingly like spilled porridge, pushing itself down and out. High above us the cliff continues to fall, crumbling in rushes and lulls. Every now and then massive thudding sections, bigger than houses, carrying with them the soil and the thorny scrub of the cliff top, plunge into the glutinous marl, shunting it ever downwards. To stand and watch this happening before your eyes is to put creation into reverse. It's on rapid rewind; instead of the land building up over millions of sedimentary years, it is dismantling in minutes, haemorrhaging back to the sea. It makes me think of that enigmatic chorus line in A.C. Ainger's famous hymn, 'when the earth shall be filled with the glory of God as the waters cover the sea'.

At the foot of these landslides the collectors sift and dig. Vigorously they shovel away debris, halting suddenly to pick a find from the morass. They inspect it minutely, turning it over in their hands, often thoughtfully tapping it with a hammer. One young man lies full length in the mire, on his side like a nineteenth-century coal miner, hacking at a vein of hard limestone with a short-handled pick, as if his life depends upon it. We ask him what he is searching for. When his mud-spattered face looks up at us we see that he has a jagged scar down the side of his jaw. He's not very friendly, avoiding our eyes, perhaps resentful of competition or disturbance. It makes us wish we hadn't asked. But he reluctantly replies, giving us one important shard of information. There are fossils and there are *fossils.*

Those laid down in young, soft rocks such as mudstone, lias or marl, although beautifully intact, are unstable and won't last. They will dry out and crumble away to nothing. They're too young. Their few million years of silent pressure have been found wanting. These are adolescent fossils, not to be taken seriously, imperfectly petrified. Beyond the knowledge their presence imparts, they are worthless, scoffed at by the cognoscenti of the fossil-collecting world.

The ones to go for, we learn, are those that have served their time – are cast permanently in stone. These are what the young man with the scarred face is after; these are what the fossil shops will pay for. It's the rock of ages he's after, not the insubstantial shells of adolescence. He is mining the cliff for his supper. Come to think of it he looks a bit lean. Perhaps it's not so surprising his friendliness is rationed. We stay just long enough to watch him gouge a fine ammonite from the rock. It is eight inches across and perfect, like a sculpture. With a chisel-headed hammer he taps away the superfluous limestone and bags it without a glance at us. Then he's straight back hacking at the seam with the fury of a Viking bent on pillage. As we walk away we can't help wondering about that scar. When the competition gets tough, perhaps they attack each other with picks.

In no time at all Hermione has found her first Charmouth fossil – also an ammonite – but one encrusted with pyrites, fool's gold, which awards it a special lustre. It is treasure; it might as well be Tutankhamen's mask. She runs to me with eyes flashing excitement, stumbling across the shingle. It will be the first of many, but the most memorable. After a while I leave her to her excavations and wander down to the sea. Sitting on the shingle I watch the breakers surge in and crash on to the foaming beach. They take over. They mesmerise. My brain begins to undulate at the ocean's bidding. It shrinks to a dungeon of emptiness. I begin to lose control, as if nothing, anywhere, matters. Hermione doesn't exist. She and the fossils and the young man with the scar have all been ground to nothing. I am reduced to utter insignificance, just another pebble, one more grain. An irrational magnetism draws me to the ceaseless energy of the waves and I have an urge to walk out into them in some sort of mindless act of sacrifice to nature, dust to dust.

Just at that moment, a few seconds before my brain liquefies and is completely washed away, a woman walks through my trance. She is stalking the waves' edge in green gumboots, head bowed, eyes scouring the fine gravels, which still hiss and foam from each withdrawing wave. She wears jeans and a heavy canvas smock with patch pockets across the front into which both hands are thrust to keep them warm. A blue woolly hat is pulled over her greying curls, giving her the wild and slightly unkempt air of a poet. She's beach-combing; for a moment I think she must be looking for shells. She stoops and picks up something small. She turns it in her fingers and then nonchalantly drops it into the pocket of her smock. Straight away her eyes return to the grainy shore. I realise that she, too, is smitten. When she has collected two or three more my curiosity can no longer contain itself. 'Excuse me. What are you finding?' She looks up, slightly startled. I don't think she knew I was there. I smile. I am relieved to see that her face is unscarred.

Perhaps deciding that I don't look like serious competition, she dips into her pocket and comes up with a handful of pyritic ammonites the size of coat buttons. They are bright gold and sparkly, quite perfectly formed in tight spiral whorls like minute Catherine wheels, ribbed and beautiful. 'Surely they're not stable?' I try out my new knowledge. Another precious secret emerges. She warms to my amateur naïvety and smiles back, mildly amused. 'Yes, that's true,' she tells me, 'but they can easily be steeped in yacht varnish and turpentine, which prevents them from breaking down and adds to their lustre. Then they are highly saleable.' She nods enthusiastically as if she too is hooked and collects professionally. She goes on, 'The best way to find them is to locate a run of pyritic sand streaming down the beach away from a mudslide. Where the waves wash through the mud you can find thousands of ammonites. But they don't last long,' she quickly adds. 'You have to find them within a few days of a mudslide, before the waves pound them to nothing.' I thank her warmly and crunch back across the shingle to Hermione, who has gleaned a weighty bagful of organic history at the cliff's foot. I get left with the heavy bag while she runs happily to the waves' edge to see what she can find. In no time at all she has a pocketful of her own.

Back at the hotel in Lyme Regis we pick over the day's hoard like burglars sorting and sharing their spoils. We have found many different shaped and sized ammonites; a dozen or more belemnites, long and pointed like marlin spikes; two slightly damaged sea urchins, micraster or holaster (we aren't quite sure which); some grainy fossilised wood and a good number of lumps of flint that turn out to be – well, lumps of flint. We wash them in the bath. We have bought a book, a simple field guide to fossils and minerals. Hermione sits spread-eagled on the floor poring over her guide and fingering the spoils of her first day. Carefully she compares them with the illustrations in the guide. She struggles with the text, trying to pronounce the complicated words: cretaceous, mesozoic, oolitic, carboniferous . . . ignoring the more specific ones in italics. 'What do you like best?' I ask.

'My ammonite,' she replies emphasising 'my'.

'Which one?' She holds up her first gold-encrusted find. 'Why that one?' I enquire.

''Cos it's my first, and anyway, it's . . .' she pauses while she dives into the pages of her guide, '. . . a hundred and eighty million years old.' There's another pause while she examines it. 'I can't imagine being that old,' she adds quietly.

'I think it's pretty amazing that it's survived all that time, don't you?' No answer beyond an absorbed nod, as her hand closes around its intricate form.

'Are we the first to see it?' she eventually asks instead, opening her fingers and revealing it on the palm of her hand as though to re-enact the discovery.

'Yes. Definitely, but not we, it was *you* who found it.'

'Oh yes,' she says, enclosing it in her fist again. 'Was I the first person in the whole world to see it?'

'Yes, of course. People weren't around when it got buried all those millions of yonks ago. Humans hadn't been invented.'

'Cool!' She throws me a broad smile.

In the morning we explore the famous Lyme Cobb. This great arm of ancient masonry reaches out into the bay to protect the harbour from the tumultuous seas that lurch up the English Channel from the Atlantic. It is constructed of huge blocks of limestone reminiscent of the ruined fortifications of some grand and ancient civilisation. The seaward wall is massive and tilted to throw back the waves that smash into it; this shelters a broad walkway along the harbour, which ends in a series of stone buildings once occupied by harbourmasters, excise men and the fishermen who traded and smuggled to and from French ports in past centuries. Over time these appealingly ramshackle buildings have haphazardly expanded on to one another in random lean-tos and low, corrugated iron-roofed shacks.

The black-painted door to one of these stood wide open, inviting nosy strangers to peer in, which is exactly what we did. When our eyes had adjusted to the darkness, we saw that it contained all the paraphernalia of a fisherman's store. It reeked of creosote and tar, and fish, long-rotted. There were stacks of wire lobster creels and barnacle-encrusted ropes in coiled heaps, dotted with brightly coloured plastic floats. An old anchor rusted silently in a corner, surrounded by drums of fuel oil and wicker and plastic crab baskets stacked inside each other. From the beams of the roof dangled orange and blue ropes, old jute macramé fenders bunched like coconuts, with dirty crimson and white mooring buoys and steel traces sufficient, it seemed, to rig out a sloop.

In the middle of all this maritime junk, on an upturned plastic dustbin, sat a small man in a scruffy jersey and rubber boots. We drew back in surprise, not expecting anyone to be in there, but his immediate smile put us at ease. He was rolling himself a cigarette. His face was swarthy and grizzled, unshaven and deceptively ageless – anywhere between forty and sixty – but you knew instinctively that whatever you stabbed at would be wrong. He blended perfectly with the tarry, shabby contents of his shack – perfectly, that is, until he spoke. His voice shattered the image like a dropped glass. It was a voice more suited to the dingy labyrinths of a second-hand bookshop in the cathedral close of a university city than to a fisherman's shack. It was a shock, the more so for its unreserved friendliness. 'Hullo there. You on holiday?' No soft Dorset burr, no hedgerow or haystack coloured these words, nor any hint of regional bias. This was the wrong voice for tying mackerel heads to creels, or splicing ropes with a rusty clasp knife. No throb of diesels resonated from his precise enunciation, no ancestral creak of windlass or belaying pin, only a puzzling sense of paradox. As if to clinch the contradiction, he fixed us with arrestingly pale eyes, the eyes of a German gundog, close to the colour of the masonry of the Cobb – eyes that might have been disconcerting had it not been for his easy smile.

In a box nailed to the inside of the shed door lay a dozen or so little polythene bags containing mostly button ammonites, but also several other fossils and nuggets of iron pyrites and assorted minerals. A crudely handwritten card said '£2.50 a bag'. Divining her interest, the man invited Hermione to rummage in the bags. 'Go on,' he urged, 'see what you can find.'

'Can I?' she asked eagerly, not quite believing her luck. She spilled the bags on the quay to pick over their contents. I engaged the pale eyes in willing conversation. What issued was a gentle philosophy of regret: too many tourists forcing locals out; fishing abandoned for pleasure trips; spiralling rents and commercial rates; house prices inflated *ad absurdum* by wealthy, retired incomers; local community life in rapid decline. Then he looked wistful, drew on his thin cigarette and changed the subject back to fossils. He asked where we had been. He proffered advice about the Lyme beaches, helpfully moving outside the shed and pointing to the cliffs. He urged us to go west from the Cobb where, he assured us, if we were quick and caught the ebb tide, we would find limestone slabs in which impressive ammonite beds are exposed at low water.

His eyes danced to the melody of his eloquence. I told him how cheap his little bags of fossils were compared with the fossil shops. He smiled again and raised his eyebrows without answering. Profit didn't seem to count for much in his philosophical reckoning. I longed to know more about him, but I had no wish to pry nor exploit his considerable kindness. He encouraged Hermione to pick the best fossils from each bag, making up two bags of greatly enhanced value. She unzipped her purse, happily paying out five pounds for her little windfall of treasure. We left feeling that he really wanted her to have them for nothing.

We toiled west along an unattractive litter- and weed-strewn beach edged by dripping cliffs. It was as though Lyme Regis had ended back there with the Cobb and the pale-eyed man. Once we were past the lines of yachts and pleasure boats on their winter stances, and a row of tightly battened beach huts, we seemed to have entered a no man's land where tourists never ventured and nothing therefore mattered. Plastic bottles and chunks of Styrofoam blew about the tide line. Like stranded jellyfish, polythene bags lay soggily in rock pools. As if to reinforce the desolation, the cliffs delivered a spiteful rain of mud and debris that splattered so persistently we thought it too dangerous to explore near them.

On a falling tide the waves here were much smaller – opaque and listless, with none of the authority of Chesil Beach. As they withdrew we realised that they were draining off a plateau of dark flat rock, bisected by occasional straight cracks. This table proved to be neither slippery nor hazardous and made much easier walking than the stony beach. A weak sun reassured us. At our feet we gradually became aware that the entire surface of the rock was covered in small white ammonites. Smothered would be a better word. There were places where it was impossible to put a hand between them; my boot would commonly cover three or four at a time. When the last inch of water had slid away, Hermione tried to chisel some fossils out for her collection. Just as the scar-faced young man had warned us, they were uncollectable. The dark mudstone split and layered easily, but immediately crumbled to a soft cheese-like paste. We abandoned it and walked on towards Devonshire Head.

We were now a mile from Lyme and were beginning to wonder whether our friend had misled us. Here the beach had become a boulder field of jumbled rocks of every size and shape. Progress was slow and cautious. We were so busy watching our feet we scarcely looked ahead. Quite unexpectedly we arrived at a two-foot sill of smooth, dove-grey limestone, the vertical edge of another flat table that stretched away from us for many yards. It was a welcome seat and we rested there a while, looking back to the Cobb in the distance. But not for long – Hermione was impatient, up and running across the plateau, intent on her exploration. She had only gone a few yards when she called out, 'Daddy! Daddy! Come and see this!' I knew she had found a fossil. I was right.

Once you have witnessed them, ammonites crowded together

become a standard against which all other fossils are judged. Those limestone beds at Lyme Regis are so startling, so unimagined after the tiny, fragile mud buttons at Charmouth and even those patterning the lias shelf half a mile back towards the Cobb, their silent presence so dramatic, that they become indelibly stamped upon the brain like a great painting or a well-loved face. Hermione was dumbstruck. She ran from one great whorl to the next even greater one, only a few feet away. No words came, no exclamations, just the rapid intake of breath and the radiant face of an unexpected prize-winner, flushed and astonished. Her journey from the tiny discovery at Eathie to this colossal ammonitic extravaganza was suddenly too rapid – too much to take in.

We stood in the centre of a broad platform of smooth, flat rock. It stretched for forty yards in every direction, disappearing into the cliff at the top of the beach and tilting gently to the sea below. It was grey and almost level, polished smooth by centuries of waves; locally it was known as 'The Slabs'. Its entire surface was studded with huge ammonite spirals like enormous draughts on an unchequered board. Some fossils were up to five feet in diameter, others a little smaller. Their ribbed shells stood proud of the surface by a few fractions of an inch so that a hand run across them could read their coils like giants' Braille. The rock was hard; too hard to cut with a hammer and chisel. These fossils are quite safe. Their sheer size demands that they cannot be removed. Here they lie, as they have for 180 million years, until the waves cut back the cliff above them and expose them once again to the sea and the sky. Their time has come again. Now, for a finger-snap of geological time, they are reliving the sun and the rain and the tangy salt air. Hermione was at that moment sitting in the middle of one the size of a dining room table. I stood beside her. A few thousand people will see them and, like us, marvel at them – before they are slowly eroded to nothing.

We were silent, wonder-struck, as if we had just entered the inner chamber of a great pyramid. Hugh Miller's 'love of life' now seemed far grander than anything we had ever imagined – omniscient and irrefutable. More than this, so powerful is their image that one is transported back to the beginning of their brief marine existence, to the events of the Cretaceous and Jurassic eras, between 70 and 190 million years ago. I spoke to the rock, or the sky, or a passing gull, or to no one at all, out loud, the one question that unstoppably spirals up at us from this cold, grey page of prehistoric nature. 'What in God's name were they all doing here?'

The now extinct ammonite was a sea creature in a shell very similar to its smaller, present-day descendant: the pearly nautilus, and the last of its kind. It was a primitive cephalopod mollusc, grouped with cuttle-fish, squid and the octopus, although these last modern-day cephalopods have either dispensed with their shells altogether, or hold them internally. It was a free-swimming carnivorous hunter of warm, shallow seas, coral reefs and salt lagoons, where, on its snail-like foot, holding its spiral Catherine Wheel-like shell erect, it prowled the bottom in an era of tropical climate long since slipped south to the equator.

It preyed upon prawn-like crustaceans and small fish, which it trapped with its many tentacles that emerged in a cluster from the open end of the shell. Concealed within these tentacles lurked jaws with a hard calcified beak and inward-curving teeth, long and pointed, with which to crunch its catch. In there too were primitive eyes and a tongue; a tough shield-like mantle covered its head, which, for protection, it could jam across the door when it withdrew its tentacles and retreated into its shell. It lived in the last, largest chamber of the shell, the other progressively smaller chambers gas-filled for buoyancy. Ammonites were very common and very prolific. They also enjoyed sex. This brought them together in large congregations, where, for reasons not clear, they died, tipped over on to their sides like abandoned tyres, to be filled in and silted over by the endless calciferous sediment of the warm sea.

We were surrounded by hundreds of these huge predatory molluscs. Just how *did* they come to die? What disaster befell these coiled colossi of the shallow seas that they sank together into geological history like this; alive and reproducing one minute, dead the next? Hugh Miller was right. This is his 'wonderful record of violent death falling at once, not on a few individuals, but on whole tribes. It is as if a whole argosy, old and young, convoyers and convoyed, had been wrecked at once, and sent disabled and dead to the bottom'.

Was it a storm that killed them, or a sudden change in temperature? Could volcanic ash have blocked out the sun so that a whole generation of marine creatures perished like the dinosaurs? Or

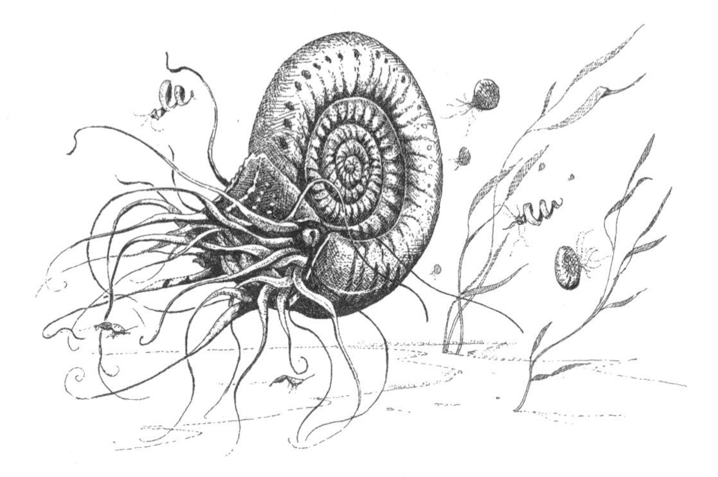

perhaps ammonites just did this: at a certain age and size they came together to breed, excited and tumescent, in countless jostling and clattering hordes, where they succumbed, exhausted and sated, littering the ocean floor with their beautiful shells. We don't know. We could only stand in silence, Hermione and I, hand in hand in their rigid midst, she with her geologist's hammer, I with a heavy bag biting into my shoulder, and marvel at this great natural spectacle, just as we had sat and pondered the mystery of Chesil Beach.

When at last we turned for home we were tired but exhilarated and triumphant. Hermione had found a single large pebble of the same grey limestone as the giants' draughtboard, the entire surface of which was a fine ten-inch ammonite, the tyre treads of its shell once more encrusted with fool's gold. It was heavy and awkward to carry, but it was booty, treasure trove; spoil to be joyously borne aloft. Its gilded spirals and ribs seemed to inscribe the strong, instinctive love of life our expedition had awarded to us both.

EIGHT

Mousa and the Simmer Dim

I only went out for a walk
and finally concluded to stay out
until sundown,
for going out
I found
I was really going in.

JOHN MUIR, 1838–1914

I had work to do in Shetland: a reconnaissance for a nature conservation expedition I had been asked to lead for Harvard Museum of Natural History. I was visiting prime nature reserves and ornithology sites, making contacts and working out logistics for the expedition. I needed to meet experts on the land, crofters and fishermen, scientists and naturalists who would address the Harvard group in the field. I needed to tease out some of the conflicts that always exist between local people and their environment. Only rarely are man and wildlife cosy bedfellows.

I then planned to visit with Hermione the far-off cliffs of Hermaness, at the very northerly tip of the island of Unst, which, together with the guano-whitewashed pinnacles of Muckle Flugga and Out Stack, are the final outposts of the British Isles – absolutely the last craggy splash. It would be hard to find a reason for going there if it weren't for the one million seabirds that choose to breed on these wild Shetland rocks every spring: gannets, puffins, guillemots, razorbills, kittiwakes, gulls and shags, all circling, screaming and bickering for their tiny territorial crannies on rows of parallel ledges, the high-rise slums of these storm- and wave-purged cliffs.

This noisome congregation is one of the great ornithological climaxes of the northern hemisphere; not only are the huge numbers of birds a great phenomenon – one of those time-eliding experiences that brands itself into your memory for ever – but the remarkably confiding nature of these birds, their apparent fearlessness of man, gives you a glimpse of what it must have been like to be an eighteenth-century explorer arriving on islands like the Galapagos, where wildlife and man had never interacted; places where man's predatory image sparked no fear or alarm. In Shetland puffins settle beside you, a foot away, as you sit on the cliff edge, while the wind from wheeling gannets brushes your face; nesting shags hiss and shake their heads at you and razorbills eye you with detached interest as gingerly you pick your way between them.

I also had another extreme of the natural world to explore, a phenomenon associated with the brief midsummer night and the peculiar wildness of one uninhabited island; something special that I decided not to tell Hermione about until the very last moment.

Once again my long-suffering wife waved us goodbye. It's not just consent that her smile awards, but the inherent spontaneity of real giving, a grace that banishes guilt, without which we would become fugitives from our own ambitions. Not that on this occasion she couldn't have come with us – Shetland is under an hour and a half's flight from Inverness, our local airport – and most certainly she would have been welcome. But this time other responsibilities held her back, and she assured us she would draw just as much vicarious pleasure from hearing Hermione recount her adventures when she returned.

It was the end of June; Hermione had just broken up for her summer holidays. There never was any doubt that she would come with me. She had overheard me on the telephone, discussing my three-day trip with the director of the Harvard travel programme in Boston, Massachusetts, and suddenly, here we were, waving back to Lucy as we crossed the tarmac to the box-like Twin Otter plane that would waft us over the round Sutherland hills, on across the waters of the Pentland Firth, to land briefly in Orkney before lifting off again, skimming the monastic loneliness of Fair Isle, and pushing northward over the empty ocean like a migrating bird.

Shetland (locals insist you should never say the Shetlands) is an archipelago of about a hundred islands, of which only twenty are inhabited. They are the last, straggling remnants of Scotland, flung out as though we'd tried to cast them back towards Norway – which just happens to be where they came from. Shetland and Orkney belonged to Norway until 1472, when they were given to Scotland, as part of the dowry of Princess Margaret, daughter of King Christian of Denmark, Norway and Sweden, when she was married by financial arrangement (a 'deal' to cancel longstanding debt between nations) to Prince James, soon to be King James III of Scotland.

Our little aeroplane landed at Sumburgh on Mainland, and we drove north to Lerwick, the friendly Shetland capital, whose narrow stone-flagged streets and family-run shops crowd together along the harbour front. We booked into our hotel, and were given a tight little room overlooking the busy harbour. A dozen jauntily painted trawlers and crab boats jostled side by side with huge oil-support vessels bristling with mysterious technical gear. The bright orange and blue Lerwick lifeboat, which seemed to throb with gallant dash, commanded a section of the harbour all of its own. Dwarfed by modern craft, an elegant, reconstructed single-masted Viking long-ship, its gunnels strung with painted shields, rocked at a permanent mooring just off the harbour wall, no doubt to reinforce Shetland's long Norse connection in the mind of the visitor. And by lucky chance, right in front of our window, lay one of the gloriously romantic, three-masted tall ships from the Continent that ply these northern waters every summer.

After supper we retired to our room. It had been a long day and we were both tired, especially Hermione. But instead of turning in we lay on our beds and dozed while the television droned. 'What are we doing?' she asked wearily, when I told her not to go to bed.

'We're going to find some Vikings,' I said. 'Have a snooze and I'll wake you when it's time.'

'Vikings?' she quizzed, propping herself up on one elbow with a hand under her chin. 'What Vikings?'

'You'll see,' I answered. 'They only come out at midnight.' I closed my eyes.

'Hmmmph,' was all I got back. In less than a minute she was asleep.

It was 29 June, only eight days after the summer solstice. At 8 p.m. the sun still streamed across the harbour, the boats zebra-striped with the shadows of their own masts. Here, above the 60th parallel, the Shetlanders have their own term for the brief, twilit June night that lasts at most for an hour. They call it the Simmer Dim. Shortly before midnight, sliding sideways, crab-like, the sun drags its gory entrails below the sea horizon, creeping low along the northern darkness and casting an unearthly green and purple glow into the brief night sky. This is the Simmer Dim. In thick cloud it becomes dark, but on a clear night the luminous day never departs for long enough to shroud the land.

We were resting not just because we wanted to see the sun go down, to sample the Simmer Dim for ourselves – I knew Hermione would love this mystical northern spectacle – but because half an hour's drive to the south, down at the tiny crofting anchorage of Sandwick, Tom Jamieson was waiting in his boat to take us over to the little island of Mousa, which lurks a choppy sea mile off the east coast of Mainland. It was necessary reconnaissance for my Harvard alumni, but its remote, glowing darkness offered something else, another of nature's extravagant flings that I badly wanted Hermione to witness.

There has to be a special reason for getting up at 11 p.m., for pulling on warm clothes and heading off into the Shetland wilds. Still more of a reason for climbing into a small boat for a draughty crossing to an uninhabited island. Hermione knew the absolute minimum, despite her mildly irritated questions on the drive.

'Why can't we see the night sky at Lerwick?'

'Because there aren't any Vikings there,' I insisted.

'Daddy, there *aren't* any Vikings. They're all dead.'

'That's just it,' I answered, seeing my chance. 'We're going to see their ghosts. We're going to a special island where Vikings were shipwrecked and where, only at midsummer, you can see and hear their ghosts.'

'Hmmm,' she said, with more than a touch of disbelief. But I could detect a hint of uncertainty, too. We drove on in silence for a minute or two. Then she asked, 'How many Vikings?'

'Oh, I don't know. I don't think anyone knows, but there were more than sixty men on board the longship when it hit the rocks.' There was another long silence. I knew she was on my hook.

I also knew that studying the Vikings at school had fired her imagination. This was a nocturnal mystery tour and she could think of no other explanation for turning out in the night, so she fell silent. Suddenly we were there. The winding road plunged down towards the sea, rounded some sheds at the water's edge, and there was the old stone jetty. The tide lapped at the boat; a handful of people were already on board. Tom, its owner, welcomed us warmly and his sure arm guided us over the gunnel. We nosed out into the purpling gloom.

Tom Jamieson has made a living from taking people to and from Mousa for many years. By day he takes them to visit the world's best preserved 'broch', a circular, forty-foot-high Iron Age defensive tower built of close-packed drystone slabs with complex galleries and stairs in the walls; a tower which has stood as a lonely sentinel over the narrow straits between Mousa and Mainland for the past two thousand years. In the year AD 910 it was in this tall fortress, a thousand years after its skilful construction by Iron Age farmers, that the Viking raider and seaman, Björn Brynjólfson – 'the ablest of men, great in sea-faring' – is said to have hidden away with his abducted bride, Thóra Hlaðhönd (the Lace-Cuffs) – 'and great joy he took to gaze upon her'. (So I learn from Elizabeth Balneave's elegantly crafted book, *The Windswept Isles*, and from the translation by Herman Pálsson and Paul Edwards of the poetic *Egil's Saga*, written around AD 1220 – one of the most remarkable of the great Icelandic sagas – which informs so much of our own history.)

Outlawed in Norway by Thóra's furious brother, the war chief Thórir Hróaldsson, 'They sailed from the east towards Shetland in wild weather and struck their ship coming a-land on Mosey. There they bare off their cargo and fared to the burg that was there and bare thither all their wares and laid up the ship and mended that which was broke.' Elizabeth Balneaves embroiders the sketch, speculating that

> they stumbled through the threshing surf over the slippery Mousa flags. Thóra would have her wedding kist full of dresses and capes and fine pleated chemises; her 'precious things', her gold and silver rings, bracelets of twisted silver and brooches ornamented in chip carving, penannular brooches and domed oval brooches from which her container for needles, scissors, knife and keys hung on fine chains . . . other bronze bound chests with eiderdown quilts and pillows, her looms for weaving and making lace and braidwork; wooden troughs and butter kirns, iron cauldrons, kitchen utensils and drinking vessels for home brewed beer and mead . . .

They must have had stores of

> butter, dried fish and possibly meat, also dried; and honey, oats and barley, onions and cheeses, along with swords, battle-axes, shields and spears for the men, great cloaks, warm furs and long-sleeved woollen jackets. All this would have to have been carried through the tiny opening at the base of the broch by one man at a time, stooping along a paved passage four feet high, four wide and sixteen long with a door halfway which could be barred against the bitter winds or the untimely arrival of an enemy.

All this activity is hard to imagine now in the draughty, roofless shell to which Tom guides his visitors each brief Shetland summer. By day he takes them to explore the nature reserve run by the RSPB where over two hundred Arctic terns nest in the long grass, the air vibrant with their scratchy cries, or to see the four hundred breeding common seals and their pups lolling about in the rocky bays. Or perhaps to enjoy the episcopal cushions of moss campion, or the spikes of delicate blue vernal squill and rugs of pink thrift on the low cliff edges, the yellow flags in the wet hollows, or the waving bog cotton – any of the dozens of other wildflowers on the island's undulating sheep-cropped sward.

But Tom's *pièce de résistance,* his dramatic encore reserved only for the intrepid and the adventurous – in our case only five others – is to return to Mousa again. To return late in the midsummer evening, just as the sun has rolled into the sea, its after-light running along the horizon like a Pentecostal fire, to slip across the glowering night waters and land us quietly at the makeshift jetty – a far prettier landing than that of the fugitive Björn, who, with his crew of sixty men, was forced to run his longship on to the rocky beach. 'Look out for the Vikings,' I whisper to Hermione as we leave the boat. Tom's instructions are clear and concise: 'Walk slowly and carefully along the coast path to the broch. Then wait.'

We do as we are told. We have small torches, although it is still light enough to cross the jetty and make our way along the single-file path that hugs the tide line for most of the half mile to the broch. We walk with a lady in her fifties who has come all the way from Bath. She is wearing a tweed skirt; she tells us she has never done anything quite like this before. We can sense her apprehension but she seems to take comfort from Hermione's presence – what a child can do she will surely be able to manage. 'What will we see?' she asks, as though she really hasn't grasped why she has ventured out into the Shetland night.

'Vikings,' whispers Hermione with such confiding nonchalance that the lady doesn't know how to respond. She drifts off on her own.

The night is still and dry. We have chosen well, although June is always the best bet for Shetland, for birds, for flowers, for weather, for daylight – perhaps for Vikings too. But it isn't warm; as the land cools a soft sea-breeze freshens our cheeks and tingles the rims of our ears. We pass through several old drystone dykes that traverse the island, separating it off into grazing parks for the crofters who have summered their sheep here every year since its last permanent residents left the island in the nineteenth century. Their roofless farmhouse rests mournfully in a hollow behind the broch.

All along the shore, above the high-tide line of wrack and tangle, runs a boulder field in a broad stripe, some twenty yards deep. It is as though the sea has collected thousands of tons of sandstone slabs the size of dustbin lids, rounded their edges and systematically cast them up in a heaped line to fix a permanent definition between the land and the sea – which is precisely what it has achieved. As the island has slowly eroded at the ocean's edge, so the striated rock has split away into flat slabs, and savage storms have gifted them back to the shore in a piled rampart. Handy if you need to build a tower.

We arrive at the broch's dark cone. Hermione stays close. If ever there was a place likely to harbour ghosts, this is it. We peer inside, stooping low through the long tunnel entrance; we flash our torch beams around the internal galleries. It is dank and smells like a cave where rotting death and the sea have conspired to make you feel unwelcome. No hint in here of the lovely Thóra and her 'precious things', or, thankfully, of the able, sea-faring Björn and his battle-axes.

We duck out again, glad to be back in the salt air. Besides, we aren't here to see the broch – far better to take your archaeology and your Viking history in daylight. No, we are here to witness one of nature's most arcane and esoteric resonances – arguably one of the most extraordinary experiences the British Isles can deliver up to the few dozen naturalists and intrepid travellers who are sufficiently motivated to travel to this Shetland isle, to be here at just the right moment in the season; people prepared to abandon the bars and their hotel beds, to gather at Tom's boat, to cross the narrow seaway and stand in the chill dark for up to two hours. And for what? To see very little, but to hear one of the most otherworldly and unforgettable sounds that nature in the northern hemisphere can summon.

As we stand here in the failing light there is as yet nothing to hear or see except a few rowdy oystercatchers racketing between themselves down on the shore. But all round us, under our feet and in the walls of the broch, in every cranny of the drystone walls and throughout the boulder field, lurk some twelve thousand storm petrels. These tiny birds, not much bigger than a sparrow, weighing just one ounce (thirty grams), are actually members of the albatross family. They have gathered here in huge numbers and occupied their underground cavities to breed. On land they are nocturnal, a somewhat limiting restriction when the night is not much more than an hour or two long; so until darkness falls there is no show – nothing to see or hear.

But then, as the sun's burgundy afterglow stains the northern sky, as though some unseen signal has triggered its beginning, a low chuckling sound emerges from – well, from just about everywhere. It is suddenly all around us as if someone beneath our feet, deep in the walls of the broch, has issued a call to arms. Thousands of avian gurgles beckon and answer each other in a low, mysterious guttural conversation. As it gathers in intensity we realise that it is ubiquitous, surrounding us, above us in the broch and below us in the boulder field. It is a three-dimensional curtain wrapping us round, a galaxy of sound in which we orbit. The closing dim becomes a cauldron of gurgling bird-sound in which we simmer gently. Suddenly a soft wing whisks past Hermione's face, brushing her cheek like a moth. She winces and draws in to my side. Looking up, we see that the luminous, velvet sky is filled with tiny, black, flickering wings.

The storm petrel is nature's dark mystery. Its name is *Hydrobates pelagicus* – the oceanic water-walker. Shetlanders call it the *alamutie* – the oily small one. The storm petrel is the smallest seabird in the world. It shrouds itself in darkness. It is sooty black with a sepia-black eye and a white rump patch. It has a tube on its nose for distilling salt from the seawater it is forced to drink during its long oceanic life. It only comes ashore to breed. Since ancient times sailors have called it Mother Carey's chicken, a corruption of Mater Cara, Our Lady or the Mother Beloved. (Another strand of folklore tells that Mother Carey was some dreaded sea-witch capable of transmuting the souls of pirates and mutineers into the 'stormy' petrels which surrounded ships in distress, condemned to wander the oceans forever without rest.) The word petrel is thought to be a diminutive (like cockerel, pickerel and dotterel) of (Simon) Peter, who walked on the Sea of Galilee at Christ's bidding: the petrel appears to walk on the water with its dangling webbed feet as it flickers over the surface of the sea searching for its planktonic food. Its name is prefixed with 'storm' because that was the only time sailors saw the birds, flocking round lurching ships, uncannily at ease in a raging wind.

Most people never do see one. Why would they? The storm petrel is strictly nocturnal on land and anyway, it seeks out only the most remote coastal and island hideaways on which to breed. The rest of its life is spent far out at sea, either making or returning from one of the great ocean migrations of the bird world. Birds ringed on Mousa have been recovered from the southernmost tip of the Cape of Good Hope. They live in perpetual summer, leaving the craggy broch and the shoreline talus by September, following the coast, passing both sides of Britain and Ireland, into the Bay of Biscay. By November they are off Mauritania, still pushing on south over the tropical seas of the west coast of Africa. They cross the equatorial doldrums to locate the turbulent water rich in plankton brought by the cool Benguela Current that sweeps north from the Antarctic and is diverted into the south Atlantic by the great wedge of southern Africa, so creating one of the richest fishing grounds in the world. Here, the summer is well under way. They will depart these rich waters before the southern autumn, pursuing the seasons back to the very same dark cranny of their birth.

Lines on a map would indicate that these Mousa petrels and those from the Faeroes, the Lofoten islands and Iceland cover over 10,000 miles in either direction, a round trip at the very least of 20,000 miles. But that isn't how a storm petrel flies. It is built for zig-zagging, surfing the ocean's self-induced slipstream. The birds' long, narrow wings and shallow, flickering wing-beats allow them to skim a few inches above the surface of the waves. They work the ocean troughs, riding along the up-currents that traverse each rolling furrow, skimming the crest and delicately surfing the downside of the next wave. Their wavering flight and erratic feeding must mean that these tiny birds travel somewhere in the region of 50,000 miles each year – over 120 miles a day.

Like other members of the albatross family, storm petrels are thought to live for decades, perhaps for thirty years, faithfully pairing with a lifetime partner and producing one chick a year. They build no nest in the darkness of their underground caverns; the single egg is laid on a bare scrape with a few bits of dry grass carried in for no obvious purpose. In common with other petrels and shearwaters (the fulmar is the most renowned), the bird spits unpleasant fishy oil at intruders. The chick is fed on regurgitated, partly digested plankton and crustaceans. The breeding birds arrive on Mousa in early May, and are gone again by October. The adults leave the fledged young behind to make their own way to sea. Only for a few summer weeks can Tom Jamieson ply his curious nocturnal trade.

So here we are. The show we have turned out for in the summer night has commenced: no raised curtain, no fanfare, no compère other than Tom's quiet instruction – 'Wait, wait and see.' The waiting is over. This bird employs only the subtlest of shock tactics. We are being slowly engulfed by a tidal wave of low, geological sound welling up from the bowels of the earth. The simmering dim was all a trick, a ploy to get us here and immerse us in this eerie spell of dark, Orphean birdsong. 'Oh, my goodness!' exclaimed the lady from Bath, and then we have lost her in the echoing gloom.

'Look up at the broch,' I whispered to Hermione. It was coal black against a sky darker than its midsummer lightness should have allowed. A ceiling of thin cloud flickered like an early black-and-white movie. It heaved with the flickering wings of what any normal person would have thought to be bats – but there are no bats in Shetland, and we were not normal people. Our normality had fled – was at that moment bleeding away into the Shetland night. Shimmering wings swept back and forth past our faces, over our heads. Storm petrels were leaving their crevices in droves. Waves of tiny birds were darkening the darkening sky. The stout walls and galleries of the broch; the shoreline boulder field and the drystone dykes; the ruined houses and byres of this tiny island were delivering up an unstoppable harvest of birds in some deep mysterious exhalation.

There was something prehistoric about this, something older by far than Björn and Thóra, than the Iron Age farmers and their fantastic conical tower. We felt that what we were witnessing – seeing, hearing and feeling – was something that had always happened, an act of creation as old as the rocks and the tides and the slosh of the ancient sea.

Hermione leaned her back against me and I placed my hands on her shoulders. Our blank faces stared at the flickering sky. There is no expression you can wear for such a thing – blank is the best you can do. Your emotions are on hold, frozen, unable to respond. You have no choice but to stand and stare as the soft wings brush your face and flutter through your vision. Your brain is overburdened with murmuring. Not overwhelmed as it would be by the sheer volume of a rock band or a jumbo jet taking off, not like that at all – no, it is just that all your synapses are blocked by a continuous stridulation of soft sound, no room for anything else.

The aeroplane, the hotel room, the television, Tom and his boat, the lady from Bath and the other four watchers in their anoraks had all been swallowed up, had vanished with the Simmer Dim. A blanket of darkness had descended upon the whole island so that nothing on the land was distinguishable. Only the black rim of the broch was faintly visible against the sky. The moon, wherever it was, had deserted us. We could no longer see the flickering storm petrels. They didn't need visibility and neither did we. Our heads were full of their chuckling, purring song, their moth wings still fluttered in our faces and the tangy smell of oily fish was all about us.

Struggling to hang on to consciousness in this ethereal onslaught, I snapped on my little torch and led Hermione away from the broch to the edge of the boulder field. The air was fresher here, just above the shore, and we sat down on the flat, dry flags. They were grainy and inviting as only sea-polished sandstone can be. 'These are your ghosts,' I said. 'In the old days pirates and sailors used to think that if they drowned at sea their spirits would return as storm petrels.'

'Why are they so tame?' she asked.

'Because they have no fear of man. They spend their lives at sea and very rarely meet a human. And we don't hurt them, so they have no cause to fear us. Do you think they're spooky?'

'No, I think they're amazing – they're everywhere.' She lay down and pulled me down to her; we pressed our ears to the soft rock.

A few inches underground, in the dark labyrinths below us, uncountable petrels were calling. All round our faces the tiny birds fluttered in and out, faintly mocking, eerie and mystical. The whole earth murmured; an unearthly gossip of ageless utterances welded our cheeks to the rock. If there is such a thing as a God of Hosts, this must be his work. The gentle tumult resonated on bare rock walls; it vibrated through our skulls, lodged in the marrow of our bones.

The world was starting all over again. The spirit of God was having another go at Genesis. The darkness was back on the face of the deep and the earth was without form and void. The firmament was still darkness and water and somehow we had got caught up in the whole awe-full stramash. This was the evening of the first day. Brochs had not been invented; God had yet to build their builders. Aeons were yet to pass before bold Björn and lovely Thóra wrecked their longship here, he with his furs and his battleaxe, she with her loom and her penannular brooches; geological ages would lapse before their cheeses and cauldrons would be lugged across these very slabs. This was the earth man had not set foot upon; he had yet to take dominion over everything upon the earth. We were on our way back to the very beginning.

It was Tom who rescued us. I felt a hand touch my shoulder and a flash of torchlight cut across my face. 'We need to go back to the boat,' were the soft Shetland words that gave us back our names. We stood up and he led us away like hypnotees, still entranced, stumbling over the boulders, back to the path.

NINE

Death in the Afternoon

> Death hath no part of me. I am that living and fiery essence that flows in the beauty of the fields. I shine in the water, I burn in the sun and the moon and the stars. Mine is the mysterious voice of the invisible wind . . . I am life.
>
> HILDEGARDE OF BINGEN, 1098–1179

Hermione was a competent swimmer long before she took to the sea. Fear of the water is a distraction; it shutters the mind and bars the way. By the age of nine she was dauntless in the security of a pool or her home loch. We had done our best to teach hcr to respect the deep and make her fully aware of safety precautions; she was ready to move on.

There comes a time in the exploration of any subject when one needs to dive deeper, discover another medium. The land has opened its heart to the probing eye and has delivered up its great forests, its deserts and its mountain tops. The sunlit sky has surrendered the darting merlin, the curving swift and the tilting kite. The muddy ditch, the river and the loch have gifted a glimpse of the teeming mysteries of aquatic life. Hermione and I often wandered along a beach near home, picking shells and turning wrack and tangle with our toes. We envied the common terns dancing with the curling waves a few yards offshore; we sensed the pulse of the ocean. Out there, beneath the waves, in a different world of plants and animals, lay a whole new ark of discovery for adult and child. We wanted to explore the sea.

The marine world seems curiously rougher and tougher than the muddy ditch (although in reality it isn't); somehow more straightforward, less prone to human misinterpretation. Its briny medium sets it apart from us, somehow mysterious yet seductive, like clouds. It keeps its distance. Fish, molluscs and crustaceans are not cuddly. Whoever heard of a pet crab? These creatures seem to spurn the emotions, not haul them in. And the fact that we can only survive in their element for brief, mechanically assisted spells hones a new edge to adventure's blade.

The Scottish seas are never warm. They're fun to splash about in on a hot summer's day, to dip in and out for a refreshing swim, but they are not the best place for a child to embrace the disciplines of snorkelling and diving. We sought heat. The Mediterranean is the closest warm water, so we picked Malta for its limestone coves and clear seas, and the village of Xhaghra on the island of Gozo, to pull focus on the dusty little place, with its hard, metallic smell of spiced pickles – a smell that seemed to belong to Africa rather than the softer essences of Europe.

We arrived one October day at a rented apartment on the edge of the village, which possessed everything we needed without pretending to be anything it wasn't. Its façade was hot in full sun and its bedrooms faced north, so that they retained the cool of the night long into the day. Even the plumbing worked. Its veranda looked out over a low valley of orange groves and smallholdings tended by their slow-moving occupants, whose voices occasionally lifted to us on the quivering air.

The island soils were thin and alkaline, a parched quality of the limestone, which instantly siphons away into deep rock all moisture falling upon it from any source. Everything on the surface was drying out. One felt that even the oranges nestling among their dark, waxy leaves would eventually shrivel to nothing. Life was brittle. Crickets rasped dryly. Fallen leaves curled and crisped like wood shavings rustling on the hot breezes that played through the groves. Lizards rattled among them. Colourful butterflies danced airily through, mischievous, frivolous and insubstantial.

If the purpose of this mini-expedition-cum-holiday was to teach Hermione to snorkel, it got off to a bad start: there were geckos on her bedroom wall. In no time at all she had collected six of these harmless, unsuspecting reptiles and imprisoned them in a box. Then she set about catching insects for them to eat. Nature had supplied an immediate, absorbing new obsession. The wonders of the ocean now seemed distant. While Hermione indulged this latest whim, I sat and eyed through binoculars and a telescope (essential equipment for a naturalist), the Gozitan world across the other side of the valley.

Old men teetered about their little terraced plots, not seeming to achieve very much, tending to a plant here and watering another there. One man I saw sank into an old deckchair in the shade of a scrubby fig tree. It was only 10 a.m. – a little early for a siesta, I thought, although the day was heating up fast and I shared his desire to sit and do nothing. As he settled I noticed that he appeared to be holding something in his left hand. The valley was a quarter of a mile wide; it was hard to see exactly what he was up to. He sat still and I moved on.

A little while later my eye was drawn back to the old man. He had risen from his chair and hurried forward twenty yards or so down a stony path. He bent over. When he stood up he was holding a small cage. He returned to a crude shed near the chair and disappeared inside. A moment later he came out again and resumed his seat in the deckchair. I saw him fiddle with something in his left hand and then settle into stillness. Slowly I realised what I was watching. This aged Gozitan was catching birds. There was string in his left hand, which led along the path to a cage trap baited with seed. Migrating birds passed over these Mediterranean islands like clouds, landing to feed and rest before heading on south to their winter destinations. There were greenfinches, ortolans, linnets, serins and goldfinches. The old man was catching them to sell to the cage-bird trade.

You cannot be a British naturalist for forty years without knowing something about this sorry state of affairs in the Mediterranean, so it was not a surprise. With my telescope I spent the next two hours closely inspecting the many different plots and groves within my view. What did surprise me was the extent of the trapping. Every allotment seemed to have either a trap or a net. Some had mist nets carefully strung between trees where small birds would fly. In the space of an hour I watched one man take eleven birds out of his net. It was impossible to identify those he was keeping; I suspect that the lucky few I saw him release were sparrows.

Cultural values differ widely between Britain and southern Europe, particularly those involving animals and wildlife. Animal suffering is differently perceived; a donkey with split hooves and pack sores is a fact of life in rural Spain, Portugal or Greece; a criminal offence in Britain. Bird-protection societies thrive in Britain; they struggle to exist in these southern countries. Their beleaguered employees are subjected to public scorn and often suffer harassment, property sabotage and social rejection. I wasn't shocked by the discovery of this pursuit beneath our noses, but I was saddened to be witnessing it first-hand. For any bird born to the freedom of the air, to be condemned to a cage for the rest of its life for the passing pleasure of a human being is, to me, a mini-tragedy which denigrates the people concerned – catcher, trader and buyer. It is a trade I wish the world could do without. I am also aware that my views are those of a member of a well-to-do post-industrial society better able to pick and choose how I earn my living than any Gozitan peasant.

Later, higher up the hillside, we watched a man shooting swallows. It made me turn away. I am far too long wedded to the chittering, squat-faced, russet-masked acrobats in electric blue that skim in and out of the stables every summer. Their friendly insistence upon returning to the same sites in the rafters year after year – a touching human-like loyalty – and their chattering company, delight me. I long for the bulging faces of their young peering down at me from the rim of the crowded nest as I groom my horse. In the autumn they gather on the telephone wires. I talk to them, wishing them well on their long, perilous journey to the Cape. Many of them will perish over the Sahara where temperatures can soar to 58°C; or they will drown, swept too far out to sea by rogue winds.

But I do not wish to be party to their dying, least of all to their wanton death at the hand of man. For Hermione and Lucy it was deeply shocking. They were angry. 'Why can't you stop him, Daddy?' was Hermione's question, when the truth of what we were seeing finally sunk in. Why indeed? Suffering exists at every level in our imperfect world; and knowing it hurts, too.

In Gozo swallows are not shot to eat. If they are badly damaged they are discarded as we might discard a shot crow; if they are not, they are taken for taxidermy. Glass cases of stuffed birds adorn the living rooms of many Maltese houses. People vie with each other for the largest and glossiest selections of exotic species. Perhaps as I write the swallows we saw sweeping through those hot, hazy skies – those we watched crumple to the parched earth – are at this moment wired rigid, permanently skimming between the television sets and the iconic triptychs of Gozitan homes, arriving nowhere.

In Britain we shoot and eat snipe, a pretty little wader not much bigger than a pipit. The Maltese shoot and eat almost all small birds – thrushes, larks, buntings – as do the Italians, Spanish, Portuguese and many other Europeans. Life for life, bedraggled corpse for bedraggled corpse, lying in a bloody tangle of broken feathers in the palm of your hand, there is little difference between a snipe and a lark. I have shot snipe and woodcock, as well as most other quarry species on the British list, although I no longer choose to do so. I have enjoyed eating them, too. The difference is another complicated abstract. How many is too many? What has size got to do with it? Is not a swallow's life as precious or expendable as that of a swan? If ethics really exist in man's habitual exploitation of nature, they are inconsistent, mercurial, and always geographical – one man's meat.

Hearing the stutter of distant shooting in the hills above the village one evening, I took off on my own in our battered hire car to explore. The dirt tracks between farms and groves snaked through hairpins winding up into the limestone cliffs. At the top of the island I was amazed to find men in full combat camouflage crouching behind every large rock and thorn. Many had dogs: spaniels, retrievers, pointers and setters. They clutched pump-action shotguns; their game bags bulged with dead birds.

As I arrived a wave of small birds appeared overhead, threading through the vermilion-streaked sunset in loose flocks, just as they do in Scotland each autumn when the Arctic chill washes hundreds of thousands of Scandinavian thrushes over our hills like hailstorms. I raised my binoculars to see the familiar undulating flight of red-wings and fieldfares – probably Russian rather than Scandinavian – streaming south to Africa. Many never made it. The air was rent by the barrage of gunfire stabbing viciously into the darkening sky. Thrushes tumbled to the rocks and the dogs were unleashed to find them and bring them in. I walked forward to speak to the nearest shooter. He turned to me smiling broadly, the success of his smoking barrels written all over his face. I asked him what he was going to do with the birds he had killed. As soon as he realised I was British he turned sharply away; he snatched up his game bag and indignantly stalked off to another position, mouthing a string of Maltese invective redolent with scorn. I was pleased that I had not exposed Hermione to this.

Malta is the only landfall between Sicily and Libya on the North African coast. This unique position attracts migrating birds of many species to roost, feed and rest – some ten million of them every year. Of these, three million are shot. Many more are wounded and die out at sea. It is a dark blot on the face of Europe, a black hole into which some of our most cherished birds disappear. Scottish ospreys regularly fall to Maltese guns; as do marsh and hen harriers, honey buzzards, falcons, bee-eaters, hoopoes, orioles and countless hordes of small birds, right down to the tiniest warblers. Despite protection laws and a very active Maltese agency desperately trying to stem this cultural cancer, the 30 per cent toll continues year on year.

At the turn of the millennium the population of Malta (including Gozo) was about 353,000: that of the European Union over 350 million. The EU has passed laws protecting most of these bird species and the habitats in which they live and breed. It is a remarkable illustration of the ineffectiveness of this pan-European collectivist legislation, and of the extent to which modern man has marginalised himself from nature, that Malta's 1 per cent – a tiny island population – can exact a 30 per cent toll over the formally identified interests of so many.

At the time of writing Malta seeks to join the EU. My head tells me to draw the islands in as quickly as possible, in the hope that the laws that govern the rest of us might at last be made to take root. But my heart damns them to eternal isolation. Suddenly these unplanned discoveries were threatening the very purpose of our visit to Gozo. The man was still shooting swallows, but we agreed not to discuss it further, at least for now. We averted our gaze, Lucy, Hermione and I, unable to comprehend whatever pleasure he was gaining from this ignoble murder in the hot afternoon. We turned away to the sea, which, I had to keep reminding myself, was why we had come to this island of ancient civilisation.

The ocean welcomed us with light and warmth. We found a small natural harbour named Hondoq: a vertical-sided slit in the limestone, running from the open ocean a few hundred yards inland on to a tiny sandy crescent strewn with chunks of pale rock. Nowhere was this sea-filled chasm wider than about fifty yards, a sort of private lagoon designed for our particular need. A few small, gaily-painted fishing boats were pulled up on the beach. The place was almost deserted. The calm water shone a vivid ultramarine, clear to a bottom of ribbed white sand and occasional stones, where the sunlight played in shimmering highlights. This was certainly no coral paradise of dazzling tropical fish. But it was the perfect place to master the art of mask and breathing tube, of creeping along the rock edges where pink and brown weed hung like delicate lace.

Hermione had expressed mild reservations about snorkelling. She had picked up the notion that it was hazardous or complicated, or somehow beyond her ability. In the event she was instantly captivated. Her interest and enthusiasm for intimately observing marine life enabled her to master the technicalities of snorkelling almost immediately. She was amazed to find that she could lie on the surface, face down, for long periods without either sinking or having to adjust her buoyancy. 'This is brilliant Dad,' was her response – there was nothing else to say. After a few mask adjustments she was confident, keen and excited, so much so that I had to restrain her from heading out towards the open sea. The absence of current or waves enabled us to waft languorously, hand in hand, through this first learning stage. Vivid sea anemones wafted their bright tentacles at us as we floated by. Little blue crabs with long claws scuttled across the vertical rock faces and vanished into crevices. When Hermione first saw these she tried to call out to me through her snorkel, producing strange hooting grunts that disrupted her breathing rhythm, so that she had to surface to regain her composure.

Below us, some fifteen feet down, huge spike-and-knob-shelled spider crabs stalked warily over the sand, suddenly sprinting away from our ominous shadows. Flotillas of ghostly, almost translucent fish in three-dimensional formation loomed past us, their gilt-ringed eyes flashing nervously from side to side. Hermione was enchanted, as I had known she would be. At last, our maritime adventure was under way.

Later that day a party of divers in jaunty fins and wetsuits of pink, mauve, yellow and blue arrived on the little beach. They were clearly novices and their tutor sat them in a row and gave them a safety lesson in the sunshine. When they were ready they humped their oxygen bottles on to their backs, checked their watches and reversed slowly into the water. Floating on the surface Hermione and I watched them fade gently into the deeps below us, a slow-motion ballet of six harlequins miming their peculiar dance, streaming silver bubbles up to our masks. Gozo is an excellent place to dive because there are so many underwater caverns in the limestone. As we traversed the island we saw many such groups, like curious, slow-moving reptiles disappearing into the sea in their ponderous, stealthy way.

Over the next few days we progressed from the security of Hondoq to broad deserted coves of open sea. Often we had to descend to these down winding cliff paths where thronging vegetation reached out to snag us in thorns, serrated grass lances slashed at our knees, and brown burrs clung to our clothes, so ferociously hooked that it took hours to rid ourselves from their prickly grip. But once on to the wild beaches it was as though no one else existed on this crowded little island. Warblers trilled from deep thickets and mechanical cicadas grated from the wind-shy trees. Fossil-studded slabs of hot rock angled down to the sand where lizards with bright blue and silver tails basked, and wizened, leathery old skinks lazed, metal-eyed in the midday sun. The tideless sea sent merry wavelets chuckling to the shore.

In these lonely coves the offshore reefs of limestone that long ago had crashed away from the cliffs were now forested with waving kelp and brightly coloured seaweeds in red and ginger brown. Luxuriant green algae sprouted in spinach-like tufts from shady corners. Through this marine undergrowth a parade of exquisite fish silently cruised back and forth. The first to catch Hermione's attention were shoals of tiny, electric blue tiddlers, squat and about an inch long, with a minute black dorsal fin like a grave accent, lending a Gallic refinement to their diminutive grace. They swam in perfectly drilled formation right up to her mask. Their blue was so startlingly vivid, so gem-like in metallic cobalt, that our hands spontaneously reached forward to feel them. Not alarmed, but unprepared to be touched, they scattered, darting between our fingers only to regain their perfect formation a second later. Large scaly wrasse with vulgar, rubbery lips hovered in dark gullies; when they saw us they turned tail and disappeared, erecting disagreeable crests of bright orange webbed spikes. Thin, silver pipefish with elongated snouts slid past eyeing us furtively, and flabby brown pouters with inane expressions and trailing whiskers gaped at us from sombre, lugubrious faces.

For hour upon hour we wafted through these sublime canyons of discovery. Hand in hand, we communicated by a squeeze varied in intensity according to the excitement of the moment, the free hand pointing at some great fish or eel as it pulsed away from our intrusion. Occasionally we got lost. One algal glade looks very much like another. Canyons twist and turn. Darker and deeper alleyways opened up, leading us on, summoning, promising new and wilder adventures beyond. So occasionally I would stop, point upwards and surface. We allowed our legs to drop, breaking into the air, raising masks and breathing sunlight, while we gently trod water and collected our bearings. I was often amazed by how far we had travelled around outlying rocks and whole headlands, or would find myself looking out to sea where I thought the beach should be.

It was never fatigue, cold or boredom that sent us ashore, only hunger – to raid the picnic basket, which lay where Lucy sat, locked into her book, ever-patiently awaiting our return. We were experts. Together we had discovered whole new continents, conquered unscaled summits and named uncharted valleys. Hermione had entered Aladdin's cave and found treasure previously unimagined.

Only at the end of the day did the effect of this sun-washed exertion tell. I went to her room to say goodnight and sat on her bed for a few minutes. Children of her age are rarely very communicative, although they often possess a lucidity and an eloquence that dispenses with words. 'How is my expert snorkeller?' I asked.

'Thank you Daddy,' was all I was going to get, all I could have possibly wished for. Desperately trying to stay awake to watch her (now freed) geckos snatch moths drawn to the electric light on her ceiling, she soon sank helplessly into an ocean of dreams.

It was time to charter a boat. We had spied offshore islands and skerries that we knew must be worth exploring. Comino was the biggest of these, a mile and a half away. Large yachts, enticing and beguiling with their silver sails, slid majestically between Comino and Gozo. Early one morning we arrived at Mgarr harbour, the island's principal port. Enquiries at the tourist office informed us that we were unlikely to succeed. Boats had to be booked well in advance and many diving parties were out making the most of the calm weather. We turned away, disappointed. Excited groups of divers passed us, clutching their weighty paraphernalia, jostling at the quay waiting to be collected; others already in boats were laughing and joking as they sped away through the moored lines of picturesque fishing vessels. We lingered enviously, sharing their excitement, knowing a little and guessing the rest; in silence we pictured their incipient adventures.

Just then a fat, Moorish-looking Gozitan in a singlet, his black hair as thick as a gorse bush, ran up to us. A stained leather belt seemed to hold him together. 'You like ship, eh?' he boomed breathlessly.

'Yes,' we said. 'Yes please.'

He turned and bellowed out across the harbour. 'Angelo! Angelo!' A young man in a handsome motor launch with a white awning was mooring his vessel to a buoy. He looked up. An incomprehensible shouted conversation echoed around us for several seconds. There was much gesticulating, pointing at us and yelling of 'Inglis!' before Angelo agreed to unhitch his boat and bring her back to the quay.

'Angelo very angry!' the gorse bush laughed deflatingly with a convulsion of his protuberant belly. Then he walked away.

Angelo was certainly very angry, although happily not with us. He had, it seemed, been stood up. A Dutch diving party of seven had made a booking for his boat for three consecutive days. He had been out with them the day before and was expecting them again this morning. But it was not to be. They had apparently fallen out with the local diving guide who worked with Angelo, and then decided to find a cheaper boat. 'Oh dear,' we said lamely, hoping he wouldn't take it out on us. 'We are Scottish,' I added reassuringly, as if no such thing would ever happen in Scotland. The boat was far too big for our needs, but Angelo soon set aside his animosity towards the entire Dutch nation and helped us load our bags and picnic baskets on board. 'We want to go to Comino. Snorkelling. All day please.'

To Comino, snorkelling, we went. Angelo was a delight, with patrician good looks, a gold ring in one ear and his long hair tied in a little pigtail at the back. The launch was called the *Hesperides*. It might have been the *Hispaniola* – he and Long John Silver would have got along fine. His English was basic; he understood much more than he could say, yet he knew from long experience exactly what we wanted. 'Iss OK,' was his grinning stock response to any request. 'Angelo OK.' And it was. He knew we sought isolation and calm water. He understood that all we wanted to do was swim from the boat in places where fish were abundant. We slid quietly through a narrow entrance into a small, circular-shaped cove. The anchor chain snarled through its portals; the engines died. The sea was flat and whispered confidingly from the pale rock walls enclosing us. Angelo loosed an aluminium ladder to the surface of the bright lapping sea.

We had come to this cove to experience what he called 'silver bass'. An identification poster pinned to the wheelhouse wall said *Spinotta* or sea bass, *Dicentrarchus labrax*. The addition of 'silver' was possibly Angelo's invention, a romantic elaboration in this instance entirely understated. I shall never think of this common species as anything else. These bass were about the size of large mackerel, but capable of growing to much larger weights than we were to see that day. They were pure silver in colour – lucent, ethereal, but individually unspectacular. What we were not prepared for were their numbers. As soon as we immersed our masks in the balmy water we were thronged round with silver fish. They were all about us. Tens of thousands of these delicately identical fish were congregated here as if they had been summoned for a great reunion of the species. Suddenly I knew how Simon, the reluctant fisherman on the Lake of Gennesaret, must have felt when he let down his net at Christ's bidding and enclosed 'a great multitude of fishes: and their net brake . . . and they began to sink' (Luke 5:6). The vast shoals hovered and circled robotically as we wafted through them, only inches away, sometimes so thick that we could see nothing but glittering silver bass in every direction.

I wondered why they were here. Why weren't toothed predators ploughing into this piscatorial bonanza? In other oceans I would have predicted an aerial raid of plunging gannets or gulls. Dolphins or seals would have been cavorting through; one of the great balaenopteran whales, a fin, sei or a humpback, would have been shovelling them up in lorry loads at a time; or man, indeed – like Simon – in our inexorably predatory way. But here in this gentle, temperate bay there seemed to be no threat, no death, no sign of nature's ravening maw, no breaking nets, just the unquantifiable bounty of the seas concentrated for a moment in time, circling and suspended in tranquillity and sun. 'Are they often here like this?' I asked before we dived in.

'Iss OK,' Angelo nodded enthusiastically, smiling toothily. 'Good fishes, eh? Angelo OK!'

For an hour we swam around in eccentric rings, entranced by this shimmering extravaganza of Mediterranean biology. I concluded that the bass must have congregated to spawn. I could think of no other reasonable excuse for so many together within so small a compass. They were like the capelin shoals off the Newfoundland Grand Banks, or passenger pigeons, or even the now legendary 'silver darlings' – the herring gluts of inshore Scottish waters in the nineteenth century. I longed to know more, but Angelo was clearly not going to be the source of any greater knowledge. For the moment all we could do was commit this remarkable moment to memory, take one last turn through the dazzle of thronging bodies before returning to the boat.

We agreed to dive, Hermione and I, for a long last look. Fish scattered as we thrashed our fins; our arms oared downward like water beetles. The water was deep here, perhaps twenty-five feet or more. We pulled ourselves down until our ears hurt and then turned and looked up. Above us the great shoal of bass had reconvened. They had closed over our heads like curtains of silver sequins, shutting out the sun so that those near the surface burned in silhouette. It was as though the lucent sky above us was one vast fish, of which each bass was an individual silver scale. We were lost in a pearly mosaic of slowly gyrating scales. Hand in hand we drifted back to the surface. We were drunk with fish, laughing through our masks at this celebration of nature's great excess. I hauled myself up on to the deck. Lucy had already come back to the boat ahead of us; I turned to offer Hermione a hand, but she wasn't there.

Unable to tear herself away, she had not followed. For several minutes we stood at the deck-rail and watched her rolling and spinning in glittering water. Round and round she twirled, spun by the slowly wheeling column of fish beneath her, caught in a dancing net of the sun's rays, performing a ballet of her own joyous devising. She seemed oblivious to the boat and the cove, to us, Angelo and to the rest of the world. The silver bass had hypnotised her, claimed her for a moment for their own, and she spun in the weft of their mystical embrace. Over and over she rolled, hair streaming, dizzy with the enchantment of the opalescent throng teeming below her.

Later Angelo took us to a lonely Comino headland with an arc of white sand, our own Treasure Island cove. A lagoon of pale aquamarine sea lapped at this beach, where we anchored and waded ashore to spread our picnic. In a while a little boat with a buzzing outboard engine rounded the headland. It nosed into the shore a little way up the sand. A man of perhaps sixty years, bronzed and very fit, pulled on a snorkel mask and fins and slipped quietly into the lagoon. In one hand he held a short stick with a large, fearsome-looking fishhook lashed to its end. He cruised away across the surface. For some minutes we watched him swim silently around. Every now and again he would suddenly dive. It was obvious he was catching something. Our curiosity grew.

We followed at a discreet distance. The water was so clear that we could see the bottom and the weedy rocks in every detail. The man dived again. He jabbed his stick into a rock crevice and hauled out an octopus. The water instantly clouded inky black – the defence mechanism employed by these cephalopod molluscs – too little, too late. Deftly the man plucked the beak and stomach out of the flailing body and the octopus fell limp in his hands. He hitched it to his belt and cruised quietly on. We watched him catch four more in this way. They were not huge, each weighing less than two pounds. He was catching his supper – Comino calamari. It was a no-nonsense, matter-of-fact harvest lasting less than twenty minutes. Then he was gone. He smiled at us as he climbed back into his dinghy and buzzed away.

We had to have a look for ourselves. Around we cruised in circles, failing to see any octopus at all until at last Hermione came across a little pile of whelk shells on the sandy bottom beside a rock. She had been collecting shells, so she dived down the ten feet or so to pick them up. As she did so, to her astonishment a tentacle emerged from under the rock and curled round the shells, drawing them together. It gave her a fright. She surfaced quickly and spluttered to tell me about it. I came over to see. There were the shells, closer in to the rock now, but apparently just as available for collection as before. I dived. As I picked up the first shell, out came the tentacle

and snatched it out of my hand. These shells belonged to someone else – that much was quite clear.

We reconvened on the surface and talked tactics. We had no stick or hook, and anyway we had no desire actually to catch any octopus, but we were damned if we were going to be beaten. The creature had thrown down the gauntlet. It had become a game, and it was our move. I returned to the boat and asked Angelo for a piece of stiff wire. We dived together and I probed under the rock with my wire. The result was electric. A jet of black ink shot out at us, diffusing into an impenetrable cloud entirely obliterating the rock and the shells and everything else. Then we glimpsed the octopus. He was a big fellow, twice the size of those we had seen caught. He was sneaking away by the back door under cover of self-created darkness. As soon as he was round the rock he shot away, pulsing his long tentacles behind him in powerful waves to provide rapid propulsion. In a split second he was gone, vanished beneath another far larger rock.

That was that. The game was over; the octopus had effected his timely escape – but we had the shells. Our day on Comino was drawing to a close. It had been blessed; nature delivering up everything we could have hoped for. And it had been a day that had certified Hermione's competence in the sea and secured her knowledge and understanding of nature at work in the warm, gentle ocean. It had brought discovery and joy, fun and laughter in full measure. It was time to go home.

As we rounded the north-eastern tip of Comino, heading back towards Mgarr harbour, we saw a powerboat zoom past. It was a bright orange rigid-inflatable, the kind used by lifeboats and police. Angelo's face clouded. 'Iss a rescue,' he muttered gravely. The boat roared into the harbour ahead of us. As we arrived we saw it again, tied up at the jetty. There was a small throng of people on the quay; an air of trauma pervaded the sunlit harbour like fog. A man called out to Angelo from a fishing boat. He cut his engines and drifted alongside. Their conversation was brief and sombre. Angelo crossed himself.

'What is it?' I asked.

'Iss very bad.' A sense of tragedy engulfed us when we moored at the quay and climbed ashore. Angelo turned white and waxen through his weathered complexion. The faces of the silent people standing around told of horror and disbelief. Just at that moment, right there in front of us they were unloading two bodies.

The party of seven Dutch who had stood Angelo up that morning had chartered another boat and gone out to a well-known diving area where deep caverns led far into the limestone beneath the island. Their new, cheaper guide was not a local. They had entered the cave in small groups. A father and his seventeen-year-old daughter had gone down together. Side by side they had swum into a long, labyrinthine chasm. They had become separated from the others. In the dark they had become disorientated; instead of retracing their steps they had gone further in, turning this way and that, searching for a way out. They had become completely lost. They were fifty feet under the sea and lost in a dark cavern. Their air was running out – soon had run out. They both died.

The bodies were not in body bags. Side by side they lay on their backs on the quay, still in their wetsuits, the father's head turned away from his daughter as though even in death he couldn't bear to face her. Father and daughter. The girl was blonde and beautiful. Her eyes were closed as though in sleep. She bore no mark, no blemish to record the snuffing out of this young life. Her skin was as pure and white as shell sand, pellucid and shining. For a moment I was transfixed. I couldn't wrest my gaze from that angelic mask of faultless youth and beauty. Her long blonde hair lay about her as if arranged by an old master, Botticelli or Tintoretto: an intolerable perfection of tragedy. She seemed to be serenely peaceful, under a spell, rapt in a dream, floating – not a corpse, laid out on a dusty concrete harbour beside her dead father, ringed around by open-mouthed bystanders. I wanted to step forward and take her by the hand, to lift her up and carry her off to somewhere as beautiful as she was, as tranquil as the lonely beaches and reefs we had explored, somewhere untrammelled and exquisite.

We hurried away. The next day we were leaving the island. The images plagued me all night. Questions loomed up at me in the dark, churned within my gut, wrung out my crying soul. A father and a beautiful daughter: whose air ran out first? Did they die hand in hand? Did they hug each other; shine their torches into each other's faces, try to say sorry, goodbye? What desperation tore at that poor father's heart, as the unthinkable consequences of his fatally flawed judgement closed in around them? What terrible remorse scourged his flailing conscience? What guilt purged those last shivering minutes? Did he fade out, still praying that some miracle would rescue his beautiful girl, or did he pray to go quickly so that he would never know her end? Or did he know it was hopeless and fight to stay conscious to hold her to him as she slid into terminal asphyxia? Perhaps his larger frame consumed the oxygen faster so that he died some minutes before her, leaving his panic-stricken child clutching at a corpse? For just how long were they forced to suffer the certainty of their doom?

In the morning a taxi came to take us away. '*Bonjoo*,' I said to the driver, forcing a smile, '*X 'gurnata sabiha harget*.' 'Hello. What a lovely day.' But I didn't mean it. I was foolishly practising the snippets of Maltese I had picked up. I knew they sounded hollow. As we loaded our bags into the car I heard a gunshot. I looked round and stared across the little valley. The man was still shooting swallows.

TEN

'Are you looking for me?'

The poetry of the earth is never dead.

JOHN KEATS, 1795–1821

Ten of us stand in the middle of the Kalahari Desert. We are in Deception Valley. It's well named, but then the whole place has deceived us ever since we arrived here twenty-four hours ago.

In Gabarone, the capital of Botswana, a city awash with its own African brand of laissez-faire, I chartered a little green aeroplane and a tall German girl, a classically beautiful Aryan blonde, to fly it. She was so slender in her ironed silver overalls, her deportment so elegant and refined, that I wondered how on earth she would fit into the tiny cockpit. She dwarfed her South African co-pilot, making it look as though his epaulettes were the most important thing about him. But when the moment came she folded in with ease and grace, both swift and svelte, like a ballet dancer. Then, for two hours, she flew us through empty skies of such searing cobalt blue that we had to squint to look out of the windows, across a haze of indistinguishable contours far below us, towards a place named Rakops.

'Somewhere just north of the Tropic of Capricorn,' was how our pilot described it to us with a wry, slightly bemused smile, which seemed to hint that not only had she never been there, but that no one else had either. This was nearly right.

We arrived bumping through cells of empty heat so that the little plane dropped alarmingly towards the desert, before bottoming out on invisible cushions, only to find that there was nothing to be seen. Our German girl's charts and instruments told her we were in the right place. Her demoralised co-pilot pored over them with an anguished expression puckering his face. He peered vacuously out of the cockpit window. He kept glancing at his watch, finally shaking it and holding it to his ear as if it was somehow at fault. They spoke in German – or perhaps it was Afrikaans – it was hard to tell above the roar of the engines. I'm sure they didn't want us to know they were lost. But as we tipped and banked in broad crescents, three hundred feet above the sand veldt, it was clear that Rakops was nowhere to be seen.

Ten minutes later we found it, but only after a protracted sequence of the sort of gut-clenching circuits some people pay for in fairgrounds. It was easy to see why they had missed it. Rakops was a tiny, rusty lesion on the face of the empty earth, a pinpoint scab on a golden apple, or a minute darn on a vast counterpane, which barely catches the eye. No more than a dozen corrugated iron shacks cowered there apologetically, without any discernible excuse, beside a faint rutted track that tailed away into the empty desert. There was no airstrip: more Teutonic muttering in clipped monosyllables, more furtive looks and yet more circling, bumping and peering out of the windows on either side as we chased the crucifix of our own shadow. The engines roared and our wing tips seemed to inscribe sweeping loops on the empty veldt.

At last a figure appeared from one of the huts and waved both arms. We circled once more and then, with spluttering engines, we skimmed the surprisingly rich grass in a succession of skips and hops, as if our pilot couldn't make up her mind about landing, before finally settling to earth. We taxied to within a few feet of the figure, a barefoot man, in khaki shorts and a sweat-stained purple T-shirt, with bright white teeth grinning from a shining black face.

We were in the Kalahari to celebrate the millennium. It was one of those wild schemes that roll from the first moment of mad inception. A throwaway line, uttered as much in mockery of the question 'Where shall we go?' as in the belief that we might actually do it, on through a few months of frantic preparation, to being here, standing beside an aeroplane in the paprika heat of central southern Africa. Beside our pyramid of dung-green kitbags we cut lonely figures in the vast, oceanic desert. We waved limply as the little green plane droned away into the even bigger, empty sky.

With me were Lucy, my adult sons and daughter, Warwick, Hamish and Amelia, and four of our staff naturalists, Duncan Macdonald, Susan Luurtsema, Jessica Seal and Sarah Kay, who run our field-studies centre in the Highlands. And, of course, under a little white hat, binoculars round her neck, clutching her diminutive backpack, the never-to-be-left-behind girl. Nine adults and a child of nine; ten jet-lagged adventurers, most of us desk-and-bureaucracy-paled naturalists who, only a few hours before, had ducked out of the spitting sleet of a sub-Arctic Highland winter, and were now giddy with the startling reality of what we had done. It felt like we'd landed on the moon, or we were Lemuel Gulliver suddenly arriving in Lilliput, wondering what was going to happen next.

We soon found out: two Land Rovers driven with all the verve of a liberating army emerged, genie-like, out of a column of orange dust from the dirt track. These intimidating, specially modified vehicles had roared across the desert to meet us, doorless and roofless, rolled-up canvas hoods flapping in the winds of their own panache, bristling with ropes, axes, shovels and winches. Out of them jumped two men who, in most other countries, or even any half-respectable town, would immediately have been arrested on suspicion of gun-running. With faces as red as the sunset, they wore sandals, shorts and sweat-stained shirts beneath broad-brimmed hats, which looked as though they had been recovered from the Alamo. Unshaven and encrusted in an icing of cinnamon dust, Ewan Masson and his father-in-law, Nick Langton, strode forward to greet us. They were to be our minders for the start of our month-long expedition to feel the pulse of Africa, our personal adventure to locate the stark nature of real wilderness in an attempt to refocus the perspectives of our work.

This place is called Deception Valley because the heat creates a mirage in its shallow, sodaic declivity; a lake shimmering in the imagination until it thins and finally vanishes under your nose. Air heated by the hot desert gets trapped in this low strip of land and is squeezed by cooler air falling from above. The air distorts under pressure, forming a lens, a fairground hall of mirrors, through which, under normal circumstances, we should be tricked into seeing a long, cooling lake.

But this January of the new millennium, and for many weeks leading up to this climacteric (irrelevant and bogus – by any logic the real millennium should have been the turn of 2000–2001), the deception had been doubled. As if trying to spoil the show, or perhaps celebrating something of its own, the seasonal rains had broken with tradition and come early – very early. They had come in anger and with real staying power; real enough, that is, to have flooded half of southern Africa, creating havoc in Mozambique and Natal (as we would later discover), and transforming the entire Kalahari into savannah-like grassland, lush and beguiling, as far as the eye could see.

It was as if the desert itself was a mirage; as if there had been a mistake and somehow we had arrived on the wrong continent. So there was no mirage lake and no desert, only rich savannah suddenly and paradoxically teeming with life, undulating gently to a 360° horizon beneath a high, punishing sun. The sandy soils had opened their pores and exhaled a rush of water-dependent fertility, turning the word desert on its head – far more than a deception, a travesty.

Files of springbok strolled through the heat-shimmer, sleek and fat. Ruminating gemsbok, the oryx of the desert, with long, piebald faces (a device for cooling the brain), posed like heraldic supporters beneath slender, halberd horns. Quilts of wisteria-blue flowers lingered in a warm fog over the yellow-green grass. Languid cheetahs lounged in the long grass on the high rim of this exquisite place, and aristocratic lions drowsed slit-eyed in the threadbare shade of acacia thorn. For the most part these predators were unseen, but their hauteur seasoned the hot air like Apaches in the Wild West, confirmed every few minutes by the sharply raised horns of the springbok, which made jerky little spurts forward before settling to graze again, as though they were making quite sure they could still flex their springs.

On the only vehicle track through this valley, white- and black-fringed butterflies clustered in thousands at the edges of mud ruts, siphoning orange moisture through tongues rolled out like party hooters. They lifted off as we approached, parting and closing around us in curtains of printed cotton as we drove slowly through. The air hummed with strangely percussive birdsong, repetitive and rasping, like old lawnmowers, occasionally punctuated by a piercing cry. It was not what we expected.

This was Hermione's African debut. Her response to such an adventure was both predictable and confident. Her familiarity with wildlife at home in Scotland had given her a fearlessness and a thorough elementary understanding of nature, which, on the one hand we knew would serve her well in a true wilderness such as this, but which, on the other made us nervous because it needed to be attuned to the many very real hazards of such a savage place.

On that first day she sat on the roof of the Land Rover and supplied a running commentary of sharp-eyed observation, something we came to rely upon in the days ahead. It was she who first saw cheetahs; she who was the first to spot the ridiculous and lugubrious faces of red hartebeest; and she who begged to stop and take a much closer look at the butterflies. Yet as the hours passed she withdrew to the familiar perspective of her home's natural history, almost ignoring the ubiquitous clots of springbok and gemsbok in favour of the invertebrates and the minutiae of the sand veldt. When we stopped for lunch she found a sinister solitary wasp that had stung to paralysis a large, glossy-green caterpillar much larger than itself, and which it was busy hauling into its larder, a deep hole in the sand. She drew us into this drama of mini-predation at our feet and made us watch out the extraordinary military efficiency of the wasp's dealings with its pathetic, doomed captive.

Every minute was a learning exponential that, as I watched her, brought a spontaneous smile to my face. She never rested, nor hesitated to tug me off to see some new discovery she had happened across. When we found a lion kill – a springbok reduced to a tangled mass of bloody skin and horns – it was the burying beetles that fascinated her most. While the rest of us scoured the thorn shadows for the lions, she was catching grasshoppers in her hat. Because I had taken her out of school for this expedition, I insisted that she wrote up her journal every day. She took to this with total dedication; going off quietly each evening to work in the shade of a tree or a tent. She pored over the work, punctuating her text with intricate maps and diagrams of the excitements she had witnessed, and illustrating it with carefully studied sketches of the wildlife itself.

So here we all are in this paradoxical, contradictory wilderness. We are camping – not that there is any option in Deception Valley. Ewan tells us that we are very likely the only people, white or black, in an area the size of Portugal. The San people, the few remaining hunter-gatherer Bushmen of the Kalahari, are long gone from this area, squeezed out by another paradox – a lethal amalgam of modern, land-partitioning conservation politics and incursion by the cattle economy of the Ba Tswana people.

An ageing Xhosa called Lot runs our camp with three Tswana boys from the town of Maun, up-country on the northern fringe of the desert. They are well practised at safari work: smiling, discreet, proffering the limp handshake of a people to whom such greetings are incidental, speaking an English of gesture and brief, verbless phrases – a sort of lingual purée of functional nouns like 'fire' and 'dinner', with rare but helpful adjectives such as 'hot' and 'good' occasionally tossed in to spice it up. They are pros: there's no roughing it in Lot's camp, no spooning baked beans from jaggedly knife-opened cans, as Hermione and I are used to doing on our own home-grown wilderness expeditions. We are paired off in smart green tents pitched in a row, each with a canvas hand basin, a bar of soap and a clean towel. Everyone sits at table for dinner with a place mat and cutlery, a wine glass and a side plate. One of the Tswana boys throws another acacia bough on to the crackling fire so that after dinner we can sit in a circle of flickering faces and let the cloak of the African night wrap itself around us.

Darkness quickly smothers the equatorial sunset; night sounds close in under an awning of stars of mesmerising brilliance. The night stretches and widens. Crickets rasp and scrape. A bird called the repetitive lark takes up a stance in a thorn on the edge of camp and begins to chant a plainsong ditty of six syllables – 'Dee-dee-dum, dee-dee-deeee' – employing only three notes. To begin with it is intriguing; we have few nocturnally singing birds in Europe, the nightingale proving the rule. This lark is a new bird for us all, a speciality of the Kalahari, so we impale its dull brown shape on a torch-beam for a few seconds of admiration. 'Dee-dee-dum, dee-dee-deeee.' Frankly, it isn't much to look at, but I suppose if your party piece is singing in the dark, it doesn't really matter what dress you wear. Nor did we know it was going to sing *all* night, every night, repeating every few seconds like a permanently stuck gramophone record. 'Dee-dee-dum, dee-dee-deeee.' It was well named, becoming a refrain to be associated with the Kalahari for all time, as the chatter of castanets is with Spain.

We fell asleep with its three notes dotting our drowsiness; only as dawn steeled the horizon did it finally give up, leaving an empty stillness arching the lull between night and day. Like rasping crickets, it wasn't the least bit offensive, nor even a nuisance, just an ever-present serenade we no longer listened to, but which we would have missed instantly had it ceased.

On the second night we had recovered sufficiently from our travels to want to stay up late. The same lark resumed its litany and we sat with our whiskies and beers recounting the excitements of the day. Ewan tossed more acacia branches across the fire. The flames responded greedily; the wood crackled and sparks plumed and flared high into the air. Smoke stung our eyes. Lions roared in the distance; instinctively we stared out into the darkness. The night air seemed to heave with foreboding, each muted roar shutting us up in mid-sentence and causing a frisson of electricity to arc backwards and forwards between flickering faces. The voice of the male lion is both claiming and demanding. 'Whose land is this? Whose land is this? Mine! Mine! Mine!' is how old hunters recount it. You are in no doubt that it's you who is being challenged, along with everything else in earshot. To us strangers from another world it is both awesomely arresting and profoundly reassuring to know that lions are out there claiming territory and killing game, as they always have done – biological imperatives which, in Scotland, we have reserved exclusively for ourselves ever since we exterminated the wolf and the lynx. It is good to be chastened in an unfamiliar environment, to shake out a few convictions, sweat a little at the neck, to feel the shiver running down your spine.

Slowly we drift back into talk. Finding adult badinage tedious,

Hermione takes off with her torch to visit the kitchen camp and watch the Tswana boys washing their pots. Nick recounts yarns of old colonial lion hunters and narrow escapes, and modern tales of lions and hyenas coming into camps and dragging hapless safari guests from their beds. Lucy looks apprehensive. Longer than we realise has slipped by. She calls out to Hermione. No answer. I stand up and call again, louder. Still no answer. Nick, Hamish and Warwick put down their beers and rise from their chairs, rubbing smoke and mosquito repellent from their eyes. I call out for a third time. 'We'll take a look,' Nick says.

'Missy OK,' a voice speaks from the darkness. A torch flickers. Lot approaches the fire; it lights up his smiling face like a pumpkin lantern.

'Where is she?' I ask, in a voice only slightly edged.

'Missy OK,' he repeats. 'See?' I follow him the twenty yards or so to the kitchen camp where, hidden by the Land Rovers, he and the Tswana boys have a fire of their own. A Tilley lamp hisses energetically from a low branch. There she is, immortal and invulnerable, on her hands and knees busily catching moths as they are sucked irresistibly into the bright yellow light. 'Are you looking for me?' she asks, scarcely glancing up.

Hitting the hot glass shade the moths fall, dazed and singed, on to the bare earth beneath. On the ground an array of upturned wine glasses encircle the lamp. Beneath each one is a captive moth, compound eyes blazing with a demonic glare. Proudly she shows me. Some have pale silver wings with huge unblinking mock-eyes of electric blue; there are big cinnamon moths with crenellated wing edges and palm-branched antennae, furry, and quivering; others have small, sleek wings of burned cork and purple, silvered with moon dust. Then she takes me by the hand to see her *pièce de résistance*. In a tin billycan lent to her by the Tswana boys, she has imprisoned a small collection of the most fearsome-looking scorpions I have ever seen.

Scorpions are common in the Kalahari. She has caught two species, a large black one and a smaller, brown one. Both are among the most menacing creatures God created. He really went to town on these. He took a spider – which is what a scorpion really is – but distorted it beyond the wildest nightmares of arachnophobes. The

scorpion's front legs are converted into serrated snapping pincers like a lobster, and the remaining four pairs of legs are armoured and clawed, allowing it to climb and scuttle like a crab. Its head and thorax are fused, stocky and rigid. Hidden beneath a tough carapacial head-shield, a pair of forward-looking eyes glower in the centre, and, as if that were not enough, up to five pairs of lateral eyes are positioned at its front corners so that it has 200-degree vision. Its abdomen is made up of twelve segments of which the last five, articulated and finely tapered, are curled over its back in a sinister 'tail'. The last segment of this arched lance bears a bulb-shaped venom sac and a sharply hooked thorn-like sting called an *aculeus*, with which to stab home its poison.

One might imagine that this was sufficient armoury to cope with survival, even in the Kalahari. But scorpions have yet another trick of darkly malevolent invention. On their undersides they possess a pair of special sense organs all their own. These organs of fiendish invention are called pectines: teeth-like chemo-receptors, which constantly test and sense the temperature and the textures of surfaces they walk over, assisting in detecting prey and sensing pheromones from scorpions of the opposite sex. The dual imperatives of predatory cunning and sexual detection are combined to make the creature a doubly effective, rapacious, heat-seeking, reproductively charged monster.

Scorpions are nocturnal, preying on spiders, centipedes, other scorpions (that figures), small lizards, snakes and mice, all of which they grip with their pincers and paralyse or kill with their stings. The underslung pectines are 'sniffing' and feeling; hairs on the mouth-parts are sensing vibrations; even the tips of the legs have minute organs for detecting movement. Some species can grow huge – up to nine inches long – and live for up to fifteen years. The creatures are crawling atrocities. I find myself thanking God they can't fly. 'How did you catch them?' I ask incredulously.

'It's easy.' She smiles up at me as though we are talking about catching newts at home. 'I dazzle them with my torch, pop a glass over them and that's that! I slide the cover of my notebook underneath. Then I turn them upside down. They can't cope with the glass.'

'You *do* know how nasty their sting is, don't you?' I try to sound severe.

'Oh yes,' she says dismissively, 'but they're quite sweet really.'

Sweet or not, I persuade her that they have to be released. We examine them closely with our torches, testing their reflexes with a grass stem. To begin with they react by gripping the stem with their pincers, the tail curling savagely forward in a stabbing attempt to sting, but quickly realising they're being wound up, they sulk and refuse to play, as though they are saving up their venom for the real opportunity their devilish sensory organs will reveal. Seizing my chance, we carry the billycan to thc edge of camp. 'It really wouldn't be very kind to release them near the tents, now would it?'

'I suppose not,' Hermione agrees, mostly because she has to. Though she is sharing a tent with her sister and fellow naturalist, Amelia, I know there will be insurrection if I allow her to take them into the tent, even in a sealed jar – everyone has limits. We tip them out one by one and spotlight them in torch-beams as, crab-like, pincers held erect and threatening, they scuttle backwards and sideways into the scrub.

We are standing alone. These are some of the moments I treasure most. By this time Hermione and I have shared four busy years of natural-history-besotted companionship and fun. Without thinking about it we have arrived at a norm – that we would explore the world's wildlife together whenever possible. This expedition is, of course, special; a celebration of the millennium by no means planned exclusively for her. But it is unavoidable that it would be special for her anyway, simply because of her age – barely nine – and the fact that

Africa and the hot desert are new to her. We are a team working together in this remarkable, emotion-tingling place. I know she is loving every minute, but inevitably there are very few moments in our crowded days when I can be alone with my daughter. I now took her hand firmly and led her out into the desert night.

We walked only fifty yards from camp, along the slender, twin-tyre-tread sandy track, the only human imprint visible on countless square miles of thorn scrub and tall grass on either side. 'Turn your torch off for a minute,' I whispered. Behind us the camp might as well have been a mile away. We could see no light from the fires or the Tilley lamp; no voices floated to us on the hot, static air. Everything was still, but the night hummed with silent presence, the true, loaded silence of all wild places, the incorruptible and abiding throb of nature. All we could hear was the solar wind, the roaring star-song of the desert night, the distant crickets and the lark, 'Dee-dee-dum, dee-dee-deeee,' muted now, somewhere behind us.

There, in that mirage of a place, when day dies the naked desert expands to meet the stars. The land and the sky merge in such darkness that you no longer know which is which; where one ends and the other begins. Heat welled up at us from the earth, diffusing the rich and sensuous scents of the sand veldt so that we breathed it in like incense. Our eyes slowly adjusted. This was the nature we had come to find – one of the earth's last wild places. A place that humbles and calms; exalts the spirit, lifting off with the heat and spiralling skyward to the Milky Way.

'What do you think of deserts?' I whisper again, as quietly as I can, praying that I won't break the spell.

'Amazing!' she answers, in a voice which fails to find itself. Her hand tightens in mine.

If only all children of the industrial west could experience real wilderness like this. Not just because it would provide an uplifting experience in their lives, or because they would enjoy it, although I am entirely sure most would. But because it would provide them with a comparative base against which to measure the artificial, destructive world of human intervention in nature, and, unlike a television programme or reading a book or a magazine article, it would claim them as nature's children, an awareness that I believe would stay with them for the rest of their lives.

I could not see Hermione. I had only her small hand from which to assimilate her mood. I don't believe she was aware of any great philosophical import loading the space between us, but I know that she has remembered this silence, and the stars and the awe that involuntarily wrapped us round that night, that held us in our place. I know that the experience helped her to understand that this is how nature created the earth and the lions and the repetitive lark and the scorpions, and that not so long ago man, the San Bushman, belonged there too. And I know that she has instinctively comprehended that it is to nature's tune, not ours, that the desert dances in this spiritual, soul-scouring place.

If I had a prayer to hurl into the night sky, to loose off into the great primed emptiness of the desert darkness, it was that this rare moment would make her understand that she, too, belongs to nature; that she is a proper part of it, then and now and forever, and that she will always value wildness for wildness's sake. That starry blackness out there brings you curiously close to God – all gods and any god. It resonates with the hum of creation itself. It is wondrous and thrilling, and a little menacing. After a while the solid shapes of bushes seemed to lean in on us – more mirages – adopting form, as if about to spring. Hermione's hand tightened again. Suddenly the deception was gone. This *was* the desert, after all.

Zipped securely in our tent I could hear Lucy's breathing rhythmically sinking into deep sleep. The lark and I seemed to be the last conscious beings on the planet. I smiled drowsily at the events of the evening, turning them over and over, thinking them alive again, examining and cherishing them, imagining that if I tried hard enough I could commit them to permanent memory, like pocketing gems. I recalled the lions' roar and our moment of parental heart-thump when Hermione failed to answer our call. I thought of Lot's toothy grin and my kneeling girl with her ring of moths trapped like ants in amber. 'Are you looking for me?' I heard her ask. Suddenly I laughed. Of course, that's it! That's what the lark's six syllables were saying to us, over and over again. 'Dee-dee-dum, dee-dee-deeee – Are you looking for meeee?'

ELEVEN

The River, the Rhinoceros and the Reticulated Wood Snake

In the end, we will conserve only what we love,
We will love only what we understand,
And we will understand only what we are taught.

BABA DIOUM, SENEGALESE CONSERVATIONIST

We live beside a river and we live with assumptions. We think we know the river, its moods and its little ways – sometimes tranquil, predictable, friendly and gifting an artist's verve to the landscape of our glen; and sometimes riled, so worked up into a bullying, roaring rage far up in the mountains that by the time it reaches us it is way out of control, surging out, all brown and frothy, to vanquish the gentle meadows. It fills the valley with a low, sinister roar, with us all day and all night, like distant surf or a shell pressed to your ear.

So we think rivers are either in flood or they're not; that they are seasonal and manageable, like the rest of the tamed world in which we choose to live. We build our houses thinking we've got the river's measure – assumptions nature loves to creep up on once in a while, and flush out. In our particularly elegant neck of the Highland woods, Glen Affric had a lovely old stone bridge that was almost ludicrously hump-backed. It arched its way over the river like a cross-section of Napoleon's hat or one of those looping caterpillars that have legs at the front and back and an omega in the middle. But it was a bridge with great style and the grace of nature's own boulders, handpicked from the stream bed, emulating the rounded curves of the mountains themselves.

It must have been difficult to build, so much so that no one would have done it for fun. There had to be real purpose in hauling the road so high above the stream. In the eighteenth century, when stone roads first happened round here, Highlanders who had lived and worked beside the river for generations, who knew its moods far better than we do now, thoughtfully erected it, boulder upon boulder, elegantly curving up to a keystone in the centre of the arch; Highlanders who didn't mind having to trail their ponies and carts up the short slope and over the bridge. 'Better to have a bridge,' they said.

For two hundred years they climbed this slope. Then, in the 1960s, Glen Affric was taken over by the Forestry Commission. They didn't like the old hump-backed bridge because it didn't suit their long, flatulent lorries. So they knocked it down. With steel and concrete they built a low bridge, flat and broad, which allowed their lorries to zip across the river, scarcely breaking wind at all. Their surveyors came with slide-rules and levels and clipboards and said, 'It will be fine.' They thought they knew about rivers. Then, in March 1966, it rained. A warm west wind from the Atlantic tore back the mountains' winter bedclothes and it rained. It rained and rained. The snow peeled back and an outrage of melt water and rain hurled itself down the river. Away went the new bridge, ripped out in the night and spread down the riverbed in shattered bits. The old men of the glen chuckled with the subsiding torrent. 'They thought they knew the river,' they said among themselves, shaking their heads.

In the kingdom of Swaziland, that little country only two thirds the size of Wales, landlocked between South Africa's Drakensberg Mountains and Mozambique, we found ourselves camped beside a river on the last leg of our millennial expedition. It was the Lusutfu River, cutting a dash from the mountains to the Indian Ocean. It was, to say the least, swollen. The rains had come early, the Kalahari was green, saltpans jostled with broad swipes of pink flamingos, and all rivers in southern Africa were tumescent and brown.

The campsite was not ours. It was a permanent camp in the wild and remote Mkaya Game Reserve established by one of those remarkable European characters forged in the melting pot of post-colonial Africa. With the enthusiastic patronage of the late king Sobhuza II (Swaziland is one of only two surviving constitutional monarchies in Africa), Ted Reilly has not only established a network of highly productive game reserves, but he has also played a major role in saving first the white and now the black rhinoceros from extinction. After thirty years of steady expansion, still enjoying the support of the present king Mswati III, Ted has handed his game parks over to his son Mick. It was the Reillys' guest camp at Mkaya that we had borrowed in the rains for the last leg of our millennial expedition.

Mick arrived in a Land Rover so ancient and dilapidated that it might well have belonged to Methuselah. He greeted us warmly and offered to take us to see the white rhinos that were part of his captive breeding project. This is a long-term project to return rhinos to the wild in many of the countries where they had been hunted into extinction for the value of their horns – an absurdity made more tragic by the ludicrous abuse of the horns by the Chinese for a spurious aphrodisiac and other groundless medicinal purposes.

The rhinos are why we had come. I first met Ted in the 1970s; now I wanted to meet his son and show some of my family and staff what an important conservation project this was. I wanted Hermione to see these great pachyderms teetering on the brink of extinction; this last prehistoric species of wild Africa, so that she could feel the earth tremble beneath their colossal bulk and perhaps one day say to her own children, 'I saw those for myself.' She needed no encouragement to read about them in advance. In fact she was ahead of the game. Every day she asked where we were going next and what we might expect to see. Then she did her homework, reading up about each exciting new species in her field guides and making notes in her journal. She became so good at this that when, one night, we happened across an unusual mammal crossing the road in our headlights near our camp, no one knew what it was except the voice from the back of the Land Rover, which coolly announced, 'White-tailed mongoose!'

It was the same with the rhinos. Although she had no concept of their bulk, knew no history of their precipitous conservation status or the stoical, indefatigable work done for so long by the Reillys against desperate odds, she knew her black from white, their lips and horns and nuchal humps, and had taken full note of their respective reputations. 'We're off to see the white rhinos,' I said as we left camp.

'Cool,' she said, beaming at me. 'At last.'

When we arrived at the boma – an extremely robust timber corral made of whole trees set close, where the rhinos are acclimatised before being shipped off to new game parks around Africa – an ostrich accosted us. This vulgar, scaly-necked, flat-footed, mannerless bird with bad breath, standing over six feet tall, strolled up and pecked Hermione in the back of the neck with a bill like a builder's trowel. It hurt. Hermione yelled, 'Daddy help!' and ran to my side, taking refuge. The bird rattled its scruffy black feathers and fidgeted its pointless, stubby wings. It came on, trying to peck Hamish and then Amelia, again and again. One of Mick's Swazi rangers ran across and rescued us, chasing it off and swearing at it in a language we were quite happy not to understand.

And there they werc. 'Oh my God!' Hermione exclaimed, hand to mouth, and then fell silent.

There is a catwalk so that you can skirt the boma high above the rhinos, ostensibly not to disturb them, but probably also for sensible safety reasons. A stench of raw manure lifts from the hot trodden earth. Fresh dung steams in fibrous mounds and the great animals mince gently round their prison, peering myopically out of piggy eyes. You get the sense that they have decided not to fight its stout poles and its huge unyielding ramparts. Their dignity is implacably on hold. But at the same time you suspect that if they were stirred into anger anything might happen, including the alarming possibility that they could barge the whole place flat, like a dam waiting to burst.

These are white rhino. They are not white at all though, just a dirty, sandy grey. Some say that 'white' is a dialectal corruption of 'wide', referring to the broad grazing mouth of the white rhino – one of the things you look for when distinguishing the two species – but then again, other experts refute that explanation, arguing that the black rhino is distinctly darker. I prefer wide; in this animal width is abundant, altogether fitting. They are much more even-tempered than their smaller black cousins, who have pointed lips like a parrot's beak for browsing thorn. Mick gives us the lowdown: which animals are destined for what project, their age and sex, their whole record. We begin to feel we know a bit about the wide, white rhino.

Afterwards he takes us to the reception area, where row upon row of rhino skulls are shelved for educational purposes. A quarter of a century of commercial poaching is recorded here in stark, sun-bleached bone, dozens of wrecked heads laid out on racks like vintage wine. None of the huge hog-like skulls have horns; all they have is a hacked stump where axes and saws have amputated the prize for their sickening trade. Mick tells us that even when his park rangers cut the horns (which are actually formed from keratin – rigid, condensed hair) from their own rhinos, in a logical attempt to prevent them being poached, the poachers still gunned them down so that they didn't have the bother of stalking a beast without a horn. We fall silent. There isn't a whole lot you can say.

The hard evidence of this skulduggery had a profound impact on Hermione, as well as several others in our team. If you have given your life to nature conservation and believe in it as a philosophy, which imbues your thinking and your actions at every level, it becomes much harder to be dispassionate, and impossible to accept the senseless, flagrant rejection of those values in favour of greed or short-term profit. It has nothing to do with beauty or sentimentality – rhinos are certainly not cute or cuddly – nor has it anything to do with animal rights, whatever those are. It is just a crushing sense of tragedy and irreconcilable loss that such an astonishing natural creation could be persecuted to the brink of extinction in such an obscene way for so ludicrous a reason.

Hermione, whose formative values inevitably reflected our own, looked to be on the point of tears. 'Why?' she asked quietly. 'How can this be happening?' It was a question none of us could answer.

On the way back to the camp in Mick's antediluvian Land Rover, out in the wilds of the fifteen-thousand-acre park, a vast bull rhino and his cow are quietly grazing the track verge where the absence of shade has made the grass long and lush. Like dinosaurs they emerge ponderously from the wet growth. Animals free and wild, albeit in a reservation, grinding cellulose into their uncertain, fragile future, recycling the rough forage of old Africa as they have done for zillions of years. Mick kills the engine.

The rhinos turn and stare, only yards away. The rain steams from their dune-like flanks. The hot smell of herbivorous sweat and turned earth envelops us. The bull rumbles like a volcano, deep and elemental. It is as though with a mighty heave the earth has somehow split open and delivered these astonishing two-ton creations to the surface, raw, geological and incredible. Fossils have stepped out of the Drakensberg foothills to force us to revisit our origins. A scene from the early Pleistocene has time-warped our day. The cow scythes the grass with her low muzzle, side nostrils blow gently from beneath corrugations of wrinkled hide, as her vast head sways from side to side. Her twin conical horns rise like curved heraldic totems

between eyes, myopic and expressionless, set in steep escarpments. Creased eye rings make her look mournfully wise. Behind large lily-shaped ears her nuchal hump rises above the hulk of her back. Plates of dense leather armour fall to her shoulders in deep folds. An impertinent oxpecker probes the crevices for ticks and lice, hitching a ride and a meal. The bowed stance of her great padded hooves belies the speed she can turn if she needs to. She has those unexpectedly fine legs and ankles of some very fat women, trim and faintly absurd. Flies clot at the dark sweat stains behind her massive shoulders.

Hermione is sitting stock still, leaning slightly forward, on the edge of her seat. We are so close there is no need for binoculars. Their presence entirely absorbs us. The expression on my daughter's face is of blank amazement tinged with fear, lips parted, eyes wide. She is aware that I am watching her but she stares past me as though, if she takes her eyes off these extraordinary mammals for even a split second, they will disappear for ever. If the expression of a ten-year-old could speak, it would seem to say, 'So this is old Africa. This is the wildness and nature that everyone has been talking about, just here, fifteen yards away, bigger than the vehicle, bigger than all eleven of us heaped together, so powerful that it could tip the Land Rover over with one toss of that great horn and those mountainous shoulders. And yet so vulnerable.'

It takes the bull several minutes to consider whether he likes us or not. Undecided, he thinks he will move along anyway. I am happy he can't comprehend what men have done to his kind – are still doing, whenever they get the chance. He leaves with his cow: their great bulks pad forward easily, almost nimbly. Our last view is two rear ends and short whiptails evaporating into the bush that closes behind them like mist. We sit in emptiness for a moment. There is no comment worth making; all known exclamations seem hopelessly inadequate. I wonder how, in the name of God's creation, a few amino acids and minerals have ended up as this incredible configuration of bizarre design. Suddenly we notice it's raining again – hard. We look to Mick to speak first. Mick Reilly, who has lived with these extraordinary beasts as man and boy; who has watched them mate and give birth, seen them hacked to pieces, spied the circling vultures, so often smelt the sickening stench of rotting flesh that brings him to yet another loss. He smiles; with the rain running in silver rivulets down his ridiculous old felt hat and pooling wetly on to his shoulders, he smiles.

We dined at trestle tables beneath a roof of reed thatch while the rains wrapped the forest in sheets of liquid gravity. The river surged past with all the focused brutality of a freight train. At intervals we walked the forty yards down to the riverbank and watched it hurtling by. We wondered how the crocodiles and hippos coped with such excoriating power. Where do they go? I wonder the same thing at home. Where do our otters take their newborn kits when a spring flash flood bruises through? Or the dippers, those dapper little red-waistcoated wrens, which scour the fast-flowing streams of the Highlands for caddis-fly larvae – where do they go to wait out the spate?

As night fell the rains eased back, as though the foot had been taken off the throttle, freewheeling, no longer driven. It was hot and refreshing; welcome wetness after the stifling humidity of the day. The snarl of the river was loud and constant, as if we were sitting beside some great factory where machinery churned all day and all night. But it had a rumbling undertone of varying pitch, suggesting that something else was going on in the darkness, something sinister we couldn't see, like secret demolition behind closed doors.

We were housed in a line of semi-permanent tents along the river valley and one or two thatched and stone-built, mosquito-netted huts on higher ground, which seemed to have been randomly dropped into the riverine forest wherever there was a bit of ground flat enough to accept a building. Between these was dense forest – jungle, really – through which muddy and slippery single-file footpaths wound and branched like veins on a leaf. So luxuriant was this forest that it perpetually thronged the paths, as though somchow they had offended Mother Nature and she had instructed every plant and vine to close them off as quickly as possible.

Of the many built camps and game lodges I have visited, Mkaya is the one that stands out; the one which appears to have been built with the greatest respect for the forest and its wildlife, and the minimum concession to its human occupants. Like many others it is basic and functional, no fuss; but it also seems to carry the pleasing assumption that those who take the trouble to go there – and that alone is quite a challenge – do so because they want to experience the ecosystem, even participate in it, rather than be cosseted observers passing through.

We had driven from Johannesburg to Mkaya in a day, trailing across the flowing grasslands of the eastern Transvaal's Mpumalanga province in our Land Rovers, and winding our way down through Swaziland over ridged and pockmarked dirt roads. It was good to stop travelling, to stretch out, take a bucket shower, ease the potholes out of our tired bodies with food, and the gritty dust from our teeth with a stiff whisky, and, nudged along by the sudden tropical darkness, to have an early night. We paired off and spread ourselves around the huts and tents.

Lucy and I had been awarded one of the stone huts up the heavily treed slope; Hermione and Amelia were billeted down by the river. I went to see them before turning in. Their tent was of a military style, spacious and robust. Although everything outside dripped, inside it was dry; the earth floor breathed a warm fragrance of jungle fertility. They had camp beds side by side. Hermione was busy collecting wildlife from around the tent; lizards seemed to be everywhere, as did large, mechanically lumbering ground beetles: an unusual abundance of life for a tent – warning signs I should have read. Just then an owl swept in through the open flap; it perched portentously on a chair as if it was about to announce something, then swept out again.

When I left I stood in the rain for several minutes. The river seemed closer. With only a slender torch-beam to guide me I followed a path in the direction of the noise. It led me back to the camp clearing where we had dined. None of the Swazi rangers or domestics was about. The camp was dark and deserted. I wandered down to the river. It had come up, there was no doubt of that, and I knew very well that if it had rained in the mountains even half as hard as we were experiencing here, there would be a flood to come. Back at home there is a delay of six or eight hours before heavy rains in the Affric hills cascade past us. I stood at the lapping edge of the water, brown as curry sauce in my torchlight, and wondered what was to come.

The torrent, the raging heart of the river was many yards away in the darkness. What lay at my feet was a gentler flood edge, an overspill of lapping water almost without current, saturating the red earth and probing into grassy hollows. 'They must know the river; it'll be OK,' I thought to myself. As I returned towards our hut I met Amelia running towards me, her torch-beam jerking wildly around. 'Come quickly,' she urged, turning back, 'there's a snake on Hermione's bedside table.'

We entered the tent quietly and carefully. Hermione was in bed, sitting upright. In the light of a candle she was talking calmly to a silvery snake coiled and apparently unperturbed on her wicker locker only a foot away from her. Its head rested on its coils and it seemed to be watching my daughter intently. It was not big – perhaps two feet long – slender with tight, shiny scales. I didn't know what it was, but I had a fair idea of what it wasn't. It wasn't a cobra, no give-away flattened hood behind its head; nor was it a mamba or a puff adder. It had no aposematic colouring, none of nature's warning flags, like the British adder's zigzag. It was coiled in repose, not aroused in anger or defence. Its linear body language didn't seem to be telling me anything, except that it was a snake escaping from the rain. Hermione looked relieved when we came in. With commendable sang-froid she said calmly, 'I'm not sure what to do.'

I folded a thick towel and eased it towards the snake. It didn't move. I was sure it wasn't about to strike at anything, just pleased to have found somewhere dry. Once I had manoeuvred a towel barrier between the snake and Hermione, I dumped the towel over its head and picked it up. I took it outside and let it go. It wound itself away into the undergrowth without a look back.

I am well aware that many people would not wish their ten-year-old daughter to sleep anywhere exposed to snakes. I also know that many ten-year-olds would not accept the presence of a snake calmly, if at all. I was proud of Hermione; that she coped is probably a direct result of her home environment. If you live and work with animals, wild and domesticated, it is possible to reach a psychological accommodation with them, a sort of karma. It's as much to do with body language and controlling emotions as with accepting the fullness of life as it comes: an indefinable cocktail of both. To overcome all outward sign of fear – panic is a killer – and replace it with a healthy respect is the first essential. Some people are temperamentally suited to that approach, others can't even begin. Some people have irrational phobias they are wholly incapable of controlling, which immediately make any situation far worse. An animal threatened or frightened may attack in defence, or, sensing fear may become far more panicky itself. The same animal not threatened will often back off to avoid a risky confrontation. Familiarity teaches by experience.

This backcountry girl who knew of toads and newts, slow worms, scorpions and jackdaws, as well as the seasonal round of births and deaths of calves and lambs, had a huge advantage. Although she had never seen a snake like this one before – a reticulated wood snake, as Mick was to tell us the next day – and was certainly not accustomed to finding one eyeballing her a few inches from her pillow, by the age of nine she had seen and handled a wide variety of animals, and she loved most of them. Consequently, and to her lasting credit, she displayed no fear, whatever misgivings may have been churning around inside her head.

Reticulated wood snakes are harmless to humans. They eat rats and mice, lizards and large insects, but not people. The snake was no threat to Hermione or anyone else; it was quite simply seeking shelter from the storm. I went off to bed and slept soundly. All I remember is the night sky distantly whitening in little flickering shocks of sheet lightning and the rain pulsing down on to our thatched roof – soundly, that is, until 3 a.m. Amelia and Hermione were standing together in the entrance to our hut, which tugged me sharply back to life. 'Can we come in?'

'What's happened?' I countered. I lit a candle. 'The river wants to come into our tent,' Amelia said stoically.

Hermione climbed into bed with her mother and I pulled on some clothes so that Amelia and I could go and rescue their belongings. I was cursing myself for being such a fool. I had seen all the signs and ignored them. My mind was too full of white rhinos and the wonderful conservation work of Mick and his Mkaya team. I had seen the lizards and snakes escaping, the lumbering beetles, the disoriented owl pursuing this food bonanza – all of which were a direct consequence of the incoming flood. These animals live with the river day in and day out for the whole of their lives. Their genetic coding has had to cope with flooding rivers for millions of years. They know the signs. Instinct tells them when to move out.

It was far worse than I had imagined. By the time we got there the river was in their tent. It was rising dramatically fast. We snatched up their bags and left. On the path we met Duncan and Jessica huddled beneath a sodden blanket, making for the shelter of the dining hut. Their tent was flooded out too, but – far worse – the tent next to them had gone. The flood had swept it away into the darkness, beds, chairs, the lot. Mercifully, it was unoccupied. We did a rushed check of the other tents and huts, but they were all higher up the slope and quite safe. There was nothing we could do but sit around and wait for dawn.

When it came, stealing silently in, followed by a muffled sun buffeted like a ship on tempestuous clouds, we were astonished to find that the water had receded. It had retreated thirty yards. I walked down to the same spot below the dining camp as I had the night before. I could see my footmarks on the mud, was able to stand in the same place. I pushed a short stick into the mud to mark the spot. It had rained all night. Judging by the lightning and the clouds, it surely must have rained in the mountains too. I didn't understand. What had happened downstream? Why had the flood receded? Had somebody opened a sluice or pulled a plug to let the torrent away faster?

One by one our crew emerged for breakfast, rubbing their eyes, yawning and stretching, each with a different tale to tell of the exciting night and the array of wildlife that had sought refuge in their tents. We discussed the river at length. It didn't seem right, this denial of physics, this tantrum of riverine rage rising like a genie in the night to take a swipe at us and then vanishing away as if it had never existed, while the rain continued to fall. We puzzled over our guava and soggy cornflakes. Warwick, who shared my bewilderment, had wandered, coffee mug in hand, down to the river to see for himself. He called out, 'It's coming up again!' We dropped our spoons and ran to the bank. Sure enough, the stick and my footprints were now underwater and the lapping edge was advancing across the trodden earth like an incoming tide. Out there the river was much the same angry brown torrent as when we arrived, continuing to issue its defiant rumble. We returned to the tables to finish our food.

Just then the river seemed to exhale. It sighed with a mighty heave of exasperation and irritation, like a huge dragon being prodded awake by some impudent knight with a spear. For a moment it roared, a long injured outcry of pain and emotion. Once again we ran to the edge just in time to see a tidal wave some six or eight feet high surge past like a stampede. Down went the water in its wake, dragging the flood back to its former level, where my footprints re-emerged, vague and silted now, but still visible, the stick swept away. Slowly we began to understand what was happening.

Our river at home meanders through man-tamed country. Fields hem it in with artificial levées, bank vegetation has been cleared for salmon fishermen, marshes and wetlands have been drained and the generation of hydro power has bridled it with concrete gullies and dams; it is for the most part thoroughly domesticated. Only very rarely, as in 1965 when it took out the Glen Affric bridge, does it go wild – a flourish of sweet, poetic revenge. But here, this African river coursing through Swaziland still possessed the quintessence of wildness, an incorruptible quality of nature which, like Mick Reilly's rhinos, once in a while broke out of the boma and surged free.

Our Highland countryside is so tidy and deforested that our river has little opportunity to collect debris. The trees it plucks from the banks are small and insignificant; the silt it carries from the empty hills is clean and fine. Dead sheep float through, and the occasional cow or red deer, but that's about it. Here, in this ancient land where man and nature are still testing each other's mettle every day, the river can raise its game, assert its ancestral right to be wild.

High in the Drakensberg Mountains it kindles its might and its kinetic authority. Feeders and tributaries channel the rain in, tugging the soil and debris from scrub forest and the quixotic, random agriculture of a pastoral people. Winding through foothills it gathers speed and force; it rips at frail banks and uproots whole trees. The river is no longer water; it has become a torrent of fluid land. Stone and timber, life and sudden death are hurled before it, churning, spinning and bowling forward in an unstoppable rush toward the lowlands. It respects no geology. It tears at marshes and rips out spits and bars; it careers over smooth rock, and claws and snatches at rough ribs of strata that traverse its course. Nowhere in this mad, hysterical, downhill slalom is the riverbed smooth and even. It loads pools and hollows with silt and stones and then sucks it all out again. It hurls great boulders forward, bruising from reach to reach; it builds ramparts and demolishes them minutes later.

Once in a while geology and debris close ranks. Somewhere in this maelstrom a stout tree lodges between boulders or sharp rocks, any rooted obstacles that protrude from the riverbed. Branches pile in. The flow slows as it is forced to surmount the obstruction. Silt and stones drop, adding mortar to the barrier. It builds and builds. The surface of the river is now saddled by a great upsurge of angry water coursing over this midstream rampart. It continues to build and spread across the riverbed. More trees catch and jam. Perversely, in its moment of rage the river is damming itself. Down there, out of sight in the liquid tornado, swirling eddies and counter-currents are piling on the flotsam. The barrier grows like a rampant cancer, invisibly spreading itself against the current.

The river is now forced to break its banks, to spread rapidly sideways to compensate for this obstruction. The flood wreaks its havoc; snakes and lizards flee, tents disappear in the night. Suddenly, unable to take the strain any longer, the original tree snaps. Water rams the breach. Boulders give in to the rush and begin to roll again. The forces have become too insistent, too relentless. With an audible lurch the whole barrier shatters and disintegrates. The river heaves and sighs. A tidal wave of water and debris purges the stream, barrelling through like a charging rhino, and the flood is sucked back into the flow behind it.

Over the space of the day we watched this quickening phenomenon again and again. The floodwaters came and went almost as miraculously as when Elisha smote the Jordan with Elijah's mantle. We were right about the overnight deluge in the mountains; by midday the torrent had moved up a gear, now scouring through at a velocity that was unnerving to watch. Nothing can have survived in that flood. The loss of life of small creatures too young, too slow or too inexperienced to move away must have been uncountable. For Hermione it was a lesson she would not forget. At a lull in this drama she tossed in an observation of her own. 'Perhaps the snake was trying to tell me something?'

TWELVE

Beavers in Norway

When we try to pick out anything in Nature,
We find it hitched to everything else in the universe.

JOHN MUIR, 1838–1914

Hermione was born in 1990 into a Scotland without beavers. There have been no wild beavers in Scotland, or for that matter anywhere in Britain, for well over 350 years. We don't actually know when they finally succumbed to extinction. It was an event apparently without even the courtesy of a file note. The last word seems to be that of Hector Boece, the first Principal of Aberdeen University, who published his encyclopaedic compendium *Scotorum Historiae* in seventeen volumes in 1527. Almost in passing Boece comments that the beavers in the vicinity of Loch Ness were hunted for their skins. By 1650 they were almost certainly gone. Fourteen generations of Scottish children spawned, born and passed on without so much as the slap of a beaver's tail.

My good friend and beaver enthusiast, Paul Ramsay, first introduced me to Songli, in Norway. He and a few other Scottish naturalists were collaborating with scientists from Nina Niku, the Norwegian agency for nature conservation, on the pan-European reintroduction of the European beaver, a remarkable success story in which nineteen countries had re-established them in their former wetland habitat. Nina Niku had generously offered us the use of a cabin in Songli called 'Haugstuggu', the high studio. This was why Hermione and I and my field-centre colleague Duncan Macdonald now found ourselves bumping up a stony forest track to this mountain hideaway. Duncan and I both needed first-hand knowledge of beavers; Hermione was desperate to see wild beavers for herself.

Her interest had started with a visit by Duncan, our Environmental Education Officer, to her little local primary school to give a talk on the European beaver, *Castor fiber*. 'They were here,' he told the children, 'and we killed them all; we hunted them out. They were good for the land and they belonged here. The land still needs them; without them it's not complete. Let's put them back again.' He struck a chord; Hermione came home asking awkward questions. Duncan had fired the children's imagination. He had asked them an awkward question of his own. 'Wouldn't you like to be the first generation of children in Scotland for nearly four hundred years to see wild beavers?'

'Yes!' they all yelled. They wanted it done the very next day.

A little dirt road signposted Songli trickled off the main highway south and west from Trondheim. We were on our way to find beavers for ourselves – to answer some questions. Norway was just the place; Norwegians have been providing answers about beavers for many years. We followed the main road as it swept along, hugging the toe of the wide and beautiful Trondheim fjord, where the sea is rimmed with cinnamon seaweed and cormorants top the weed-fringed piles like heraldic griffins. Rafts of eider duck and mergansers bob like flotsam in its tranquil bays and coves. Steep mountains seem to have trapped this long leg of the North Sea, swathed in forests of native pine and spruce, not dark and forbidding as man-planted forests are in Scotland, but in a constantly changing mosaic of light and space, which awards grandeur to every slope and living presence to each cluster of trees.

Later, when we explored these forests on foot, we spoke to their owners and found ourselves in awe of their karma; the light and space of the forest was reflected in their whole philosophical approach to nature. These Norwegian woodsmen displayed quiet confidence in their sense of empathy with their environment: an inherent understanding gained from long economic association and the fact of being a part of the forest community themselves. It is said in these parts that every Norwegian knows the land of his origins. Many leave the cities and return to spend their leisure time in a family *hytte* (literally a hut), a simple cabin, the more rustic the better, tucked away in the wild forests of their forefathers. They undertake regular ritual pilgrimages to refresh their deep association with the land and its wildlife. It is a culture that values and welcomes the beaver.

For us Songli is a happy accident. Once it was a royal hunting estate. Now the elegant timber lodge is surrounded by its own nature reserve of forests and bogs. It overlooks a large lake where black-throated divers wail, and wigeon whistle and haggle in the marshes. Cranes – birds of heaven, as Peter Matthiessen has called them – uttering wild fluting calls circle overhead, glide serenely to land and stalk its marshy edge. In *Marshland Elegy*, Aldo Leopold called them 'the symbol of our untamable past', their calls 'the trumpet in the orchestra of evolution'. Here, every spring, these ancient birds dance their high-stepping, wing-raising, bowing and neck-looping courtship with all the strutting panache of the flamenco, as if they are emulating the ritual formality of the royals and their courtiers who once frequented this enchanting place.

The lodge is now a research station run by Nina Niku. The old timber building, with its simple grandeur and bare white flagpole where once the royal standard rippled in the breeze, is now preserved as a centre for seminars and research gatherings. A hamlet of barns, workshops, timber cabins and *hyttes* has sprung up around the lodge, as mushrooms surround a venerable old stump. Forest and mountain hold this picturesque encampment close, clutched like a jewel. At night, ghostlike, beneath vast palmate antlers, bull moose loom from the woods and tiptoe between the buildings on their way down to the sweet willow browse of the lake shore, leaving only the deep indentations of their huge cleaves in the soft grass. The forest ranger and his family occupy one of these houses all year round; the rest are outposts for itinerant researchers like us, in need of temporary accommodation.

The reintroduction of the European beaver to Scotland is a hot topic in nature conservation circles and in the politics of a newly devolved Scotland. It is long past the 'Should we or shouldn't we?' stage. People have been consulted, in the way things have to be done these days, and the answer was a clear, if not resounding 'Yes' from 68 per cent of those who took the trouble to respond to the document proposing the idea, put out by our own government agency, Scottish Natural Heritage. 'If it belongs here and is a valuable member of our fauna, and we did it in, then we have a duty to replace it,' seems to be the broad public consensus.

Next comes the doubt and deliberation. When, where, how many? What if conflicts arise and what happens if the land and water the beavers colonise belong to someone who doesn't like them? Scottish Natural Heritage, whose responsibility it is to process all wildlife reintroductions, proposes a pilot project spanning ten years. Research, constant monitoring and assessment, *and* an exit strategy – whatever that may mean. A public policy based upon watching closely, documenting the findings, and perhaps ducking out again if too many conflicts arise.

Others want action. There are people in Scotland ready to take beavers on their land; people who understand the role of the beaver in a wetland ecosystem, who genuinely want the species to be back where it belongs. Its absence is entirely man-contrived; there was then, and there is now, plenty of suitable habitat for beavers in Scotland. Long before we understood the principles of ecology, we hunted them to extinction for their skins, their meat and their oily scent. Then we went on to do our damnedest with their larger, but close relative, the North American beaver, *Castor canadensis*.

By 1600 hunting in Europe was not meeting the insatiable demand for beaver fur. King Charles I urged American pioneers to 'exert themselves' in supplying beaver pelts for Britain. In 1687 the Hudson's Bay Company immortalised the animal in its heraldic seal: four beavers with fat tails squatting between two moose supporters. The beaver was the mainstay of the fur trade as the company pushed west throughout the nineteenth century. It even became a unit of currency. In 1857 brass tokens called 'made beavers' were issued to Indians and trappers in return for beaver pelts. These tokens could be exchanged at any Hudson's Bay Company trading post for goods such as beans, gunpowder, shot, coffee and tobacco. The beaver was cash.

The result was its extermination across large areas of North America. Fashionable fur coats throughout the Old World were beaver. The well-to-do wore fur hats or hats fashioned from beaver felt; in Russia they still do. Beaver fur is special. It is rich, thick, long and seductively soft. Evolved simultaneously to keep out water and contain heat in extreme climates, in air alone it becomes simply wonderful. In the nineteenth century, lawless French-Canadian lumberjacks fought and shot each other for beaver-fur vests, and in midwinter socialite ladies and gents paraded the Champs-Elysées with cosy impunity. Beavers were in.

As well as this there was the oddity of beaver castoreum, an oily and musky-smelling, creamy-brown secretion from twin glands in the genital area of both sexes, high in salicylic acid extracted from the willow and aspen bark beavers love to eat. These glands produce scent for sexual attraction. To humans castoreum was part myth and magic, confused by folklore and tradition, and part real animal by-product medicine. No one has heard of it these days; its quasi-medicinal uses as an aphrodisiac, an antispasmodic and a relief for hysteria, menstrual problems and arthritis, have long been usurped by aspirin and other proprietary drugs. Men also ate beaver meat.

The North American beaver, although depleted throughout much of its range, was too widespread and too populous to be trapped to extinction. But in Britain it was different. An expanding human population and increased mobility were to spell doom to fragmented beaver communities unable to conceal their presence in a small country. By 1700 they were a distant memory, even in the remote glens of the Highlands. The same thing was happening the length and breadth of the Old World. For two centuries the creature were clinging on only in the most remote and climatically extreme areas of the far north. By 1922 it was thought there were only 1200 animals left in Europe, spread between eight remote locations.

Just who started the reintroduction trend is not altogether clear, but it was soon to catch on. By the turn of the twenty-first century nineteen European nations had already acted, either with a full reintroduction, as in Sweden and Denmark, or as a restocking process in areas hunted out, as in Norway. Both were a great success, the essence of which was the international recognition that the beaver is, in ecological terms, a keystone species – perhaps the most important keystone species in the northern hemisphere. The animal is a geomorphological agent. It creates habitat for thousands of other species. It holds back nutrients within the wetlands it creates, preventing their rapid loss to the sea, enabling them to fuel the great web of species that freshwater wetland can support. Their ponds and lagoons lay down fertile alluvial silt, building precious soil. The habitat they create supports a wide diversity of plant and animal life that would otherwise simply not be there. Well over 100,000 beavers now exist up and down the length of Europe. The same process has worked for its North American cousin. Both beaver species must now be among the most widely reintroduced mammals in the northern hemisphere. 'Why not Scotland?' Hermione asks.

We park our car on the edge of the track and step into a wet wood. Hermione is excited. She sticks to my side, dressed in khaki. She has a little green sunhat shading her face and her binoculars round her neck. Goat willows close in around us, dissolving the human world in a pastel cloud. Happily, mankind has failed to find a use for wet woodland. (Too often we drain it just because we don't have a use for it.) Cushions of sphagnum moss and woodrush carpet our footsteps; tufts of sedges and rushes and other coarse, wet-loving grasses confuse us, so that we are never quite sure whether the ground is soggy or not. Often it is. Our route is haphazard, with frequent detours around patches of dark, open water. We walk on. The road is ten minutes behind us now. Dead wood, mostly of birch and willow, exuding soft fleshy fungus and bearded with whiskery lichens, lies across our path making our progress slow, and noisy with cracking sticks. The air hangs damp and chill, the sunlight in long piercing shafts paints tiger stripes across the yellowing grass.

We stop. This birch tree across our path is not dead. Its leaves are green and fresh and its bark bright and papery. For a moment we don't comprehend what we are seeing. The eye is drawn along the straight stem to the butt, where the ground is littered with woodchips the size of potato crisps; the stump is chiselled to a pencil point. Realisation floods in, expands into silence. Hermione picks up a woodchip and fingers it reverently. A beaver has felled this tree.

I am taken aback. I had not foreseen the power now evident in front of me. I knew, of course, that beavers felled trees. I had seen dramatic evidence of that in North America, where the separate larger species, *Castor canadensis*, is more ambitious in its felling habits. Yet to be suddenly confronted with a birch tree of about ten inches in diameter felled across my path at two feet above the ground, is bracing. The hard evidence of the efficiency of this animal's work – a tidy pile of chips encircling the stump, the fresh white cut there to touch, smooth and clean – rams the impact home. 'Look at the size of the chips,' Hermione says, handing me a chisel-shred. I pass it to Duncan. Hermione collects a handful and puts them in a polythene bag, tucking them away into her little knapsack, precious specimens for her natural-history museum at home.

The sub-culture of my own twenty-five-year-old Highland experience was momentarily confounding me. Wasn't this the tree we had fought so hard to protect in Scotland? Native birches, *Betula pubescens*, are sacred in the minds of Highland conservationists still smarting from the systematic destruction of upland native woods and moorland by the all-engulfing, blanket mono-culture of Sitka spruce plantations that raged throughout the 1970s and 1980s, an insult on top of centuries of overgrazing and burning of the Highland hills. Birchwoods are special; they create soil and fertility and are rich in wildlife.

Can this sacrilege be allowed? I wondered, looking down at the fine birch tree cut off in its prime. Slowly, logic and sanity filter through, colouring my vision. The reality, of course, is that when it's given a chance, this wild, pioneering tree grows like a weed throughout northern Europe, as it does in the Highlands of Scotland. Besides, the birch, the willow, the aspen and the beaver co-existed here and in Scotland for countless millennia without a problem, until man came along and modified everything, driving wild woodland back like an enemy. There should be nothing shocking or surprising about the felling of this tree at all. What I am witnessing here *is* wild nature; it's what is supposed to happen – it's what *did* happen in

Scotland for thousands of bountiful, creative, wetland-building years. We walk on.

Very soon the wood is wet. Very wet. Looking ahead I can see open water. We are approaching a lagoon. All along its edge the evidence of beaver activity is everywhere. There are paths radiating out into the wood from the lagoon – broad, well-worn paths where the grass is depressed. We skirt the pond. Suddenly there is the tinkle of falling water. We see the stream, a small river, snaking through the wood. As we emerge from the trees a dam confronts us, its silver-bright water flashing and spilling over the rim of a rippling weir. Behind the dam the water is dark and still; the beaver pond stretches back into the wood and out of sight. We stand in silence. So *this* is a beaver dam.

Can animals have built this? Were there no engineer's drawings? Did no helmeted surveyors with theodolites and levels do calculations here? Were there no logarithms or square roots, no tricky coefficients employed on this site? We see no site office – yet the dam holds. It works; it spills at the top, as it's meant to, and nowhere else. It has been plastered with mud; it's stitched into either bank with carefully placed boughs. It was built from the bottom up, the level of the lagoon rising around the beavers as they worked. They watched the water as it crept out and round their work; they rushed to plug the breaches with mud, then observed it rising again; finally plastering the flooded side with an entire coating of mud and clay, perpetually reinforced with new sticks, instinctively placed. We stand in silence, gobsmacked by this remarkable feat of rodent industry, of uncivil engineering.

At first Hermione can't believe it. 'What was here before?' she asks with a face of open incomprehension.

'The stream,' I answer.

'But there must have been a waterfall or something . . .'

'No, nothing. Just the stream.' Silence, while she adjusts to this seemingly impossible idea.

'Why did they build it here?' she asks, as if to allow herself more time to get used to it.

'I suppose the banks were good and strong and they just liked it,' I muse. 'But it's what's behind it that really matters to them.'

'Why?'

'Because that's what it's all about. They dam streams to create a pond backing up into the woods they need as a food supply.' We look back across the lagoon, which stretches away into the birches and willows almost as far as we can see.

The birches, alders and willows of the flooded woodland are their winter food-supply; they have extended their watery habitat right up and into the food source. The dam is a remarkable structure. First they have lodged stout sticks in the streambed, some large, many small. Then they have dragged and floated into place a mesh of small brushwood and leafy twigs that have become entangled in the sticks. Then comes the mud. Collecting it from the banks and the bottom of the river, hugging it to their chests, they have paddled up and busily pressed it into the upstream side of the twiggy tangle. This dam is fifteen yards across between the earth banks of the stream. The lagoon behind it is certainly four feet deep, probably deeper. Cautiously, not quite believing in it yet, Hermione steps out on to its wall. I follow. The accumulation of material has made a structure that is both stable and strong enough for us to walk right across its top, a yard wide.

Beavers fell trees and build dams. That is what they are known for, but there, alas, common understanding of this remarkable rodent ends. The ecological reality, however, is very different. The beaver creates habitat, not just for itself, but also for everything else that likes and needs still freshwater ponds and wetland. The list is endless: water voles and shrews; birds such as waders, ducks and herons; frogs, newts and toads; many species of fish as nurseries for hatching and rearing their young; water bugs of every description, and everything from the lissom otter to the fiery-eyed osprey, which seeks to feed upon those fish, or birds, or bugs, or the wetland plants that thrive in flooded beaver woods.

The lagoons backed up behind the dams are natural filtration and settlement ponds, rich in nutrients, primed with latent energy. Eventually, over hundreds of years, they will fill in and create alluvial soils of great natural fertility. In Norwegian the word *beverfeld* means a high alluvial field created behind a beaver dam. Mountain folk understand the value of these fields. They contain some of the best soils on their tiny valley farms.

Walking across the dam and back again was a good ploy. It has helped Hermione and Duncan understand the dedicated efficiency of these rodents' engineering ability. To say that the structure was robust would be an understatement. We bounced up and down on its middle, the weakest part, and it was unmoved. It sprang no leaks, no stick became dislodged, no mud fell away.

'How do they know what to do?' Hermione asked, still in awe of the whole undertaking, as we stood in midstream with several inches of water spilling over its top and tugging at our boots.

'Instinct. Pure instinct,' I replied. 'I'm sure they learn by trial and error as they go along, and probably pick something up from their parents and other family members, but most of it is driven by entirely instinctive behaviour refined over millions of generations.'

'Magic!' she says, in the way that children crystallise opinions in one word.

'Yes,' Duncan agrees, smiling and shaking his head with wonder. 'Magic is right. There are times when I think that the whole of creation is pretty magical.'

At dusk we return to try to see the beavers at work. The sun has drifted west to the mountains where it slides, as though embarrassed, behind the blueing forest slopes. We retrace our steps to the dam in silence, carefully avoiding the cracking of sticks underfoot. I wonder how the moose moves its great bulk so silently through these woods. Duncan and Hermione settle themselves at the foot of stout birches, two silent observers watching the dam, and I move off a little way upstream. I impress upon Hermione that she has to be very still, but I have few worries. She has watched wildlife with me before and knows the rules.

There is no wind. The stream burbles comfortingly. Behind the dark mountains the sky is on fire. Gnats dance in vertical columns among the birches. I settle only a few feet from the lagoon. Beside me a bluebottle zigzags noisily over some fresh moose droppings, fibrous and green. I resist shooing it away. Minutes blur. In a pine off to the right, a woodpecker chinks and taps like a sculptor, occasionally pausing to admire his work. I can barely see Hermione and Duncan in the gathering gloom. Small fish are skimming about beneath the surface of this beaver lagoon. I wonder what they are. Are there sticklebacks and minnows here? Or are they trout parr? I must remember to ask . . .

Ssquplaash! The water beside me erupts volcanically. A beaver has surfaced only three feet away. It has immediately seen me, smelt me, or both, and dived again with a ferocious slap of its tail on the water, alerting its friends. Rings echo outwards across the water as black as an old vinyl record. I see Hermione and Duncan craning their necks to see what the commotion was. I signal enthusiastically, pressing my finger to my lips. We wait again. Will they return now that they know humans are about? Is that it?

It is a now unfashionable Victorian ideal that patience is a virtue – somehow good for the soul. I have always found it a testing discipline, which took decades to train into place. God didn't include the correct software; I had to buy it in, expensively. I had to learn by the long route that it really does deliver. Even now I still have to fight off what I call the 'Oh, sod it!' instinct, which insists that it would be more fun to go home and put my feet up in front of the fire with a dram. Years ago, while badger-watching, I found that a good trick is to doze. Some of my most engaging wildlife sightings have been when I have settled down against a tree and nodded off. Once in a Highland pinewood I was awoken by a cock capercaillie – the huge forest grouse – pecking grit from the soles of my boots. On another occasion in Tanzania – a little disconcertingly – I awoke to find myself staring into the coldly drilling eyes of a leopard, all of ten feet away.

I am fully awake now. I am sure I heard something. The lagoon is rippling again, this time it's over by the dam, near Duncan's dim shape huddled against a tree. A sleek, wet head surfaces and silently streams across the water. I can see its long body behind the round head. A V-shaped wave issues back from its steady nose powering across the pond. It is heading for the dam. Seconds later it is there, climbing out on to the shallow weir, the dam's rim. I can see the whole animal now, the hunched shape, the shaggy fur clotted in streaks, and the broad, hairless, scaly, spatula tail. I know Hermione and Duncan can see it too. It moves along the dam attending to a branch here, a twig there. It seems to be an inspection – a maintenance check. It's OK. No need for repair, just a nudge and a snip in passing. It comes ashore at the far end of the dam. I have to crane my neck to see it now. Perhaps I shouldn't have . . . it's gone. Slipping noiselessly into the lagoon it dives and disappears. The ripples flatten. We wait again.

But the show is over. That's it. Darkness fills the wood until the lagoon reflects only a lemon segment of moon rising in a charcoal sky pricked with stars. There is no point in staying on. Our initiation is complete. We have seen the wild beaver and it has seen us. It gave us a moment of its busy, creative existence to validate our own. I feel a sense of elation, almost of transfiguration. We have rejoined the past. Duncan and the girl who had *not* seen a beaver in the wild are no longer with me – they slid into oblivion with the bleeding sun and the purple mountains. We feel as though the world is now divided into those who have and those who haven't. We smile to each other, struggling for the right words to impart our excitement, our accomplished mission.

The next day we head back to Trondheim to see beavers in man's bustling environment. We could have done it the other way round: seen them where they are accustomed to man first, to get to know the animal, and then tried our luck with the wilds. But I am glad to have done it the way we did. We return to Trondheim feeling we have earned the next step. It no longer matters what happens, from here on it is all bonus. We have met the beaver for real.

On the edge of the noisy, bustling city, in the reservoir that quenches the thirst and washes the clothes of 140,000 people, and hard alongside a golf course, is a population of beavers so familiar with human presence that they all but ignore you. Thysendammen is a reservoir surrounded by a city park. Pairs of gossiping, pram-pushing mothers circuit the tarmac paths around its perimeter, oblivious to the large aquatic rodent squatting quietly in the shallows a few feet away.

Whole classes of school children being environmentally educated stand noisily on the concrete dam, shrieking and pointing as a furry-headed bow-wave streams across the water in front of them, heading for a marshy feeding ground on the other side of the lake. Golfers searching for lost balls in the sedgy shallows barely comment when a shaggy animal the size of a spaniel slides out into the water and dives in a ring of light-reflecting ripples.

Our own experience is no different. After only a few minutes of sitting quietly in the haemorrhaging sunlight, we see the beavers emerge from their lodge, a clearly visible hump of mud and sticks on the far side of the lake. They appear one by one, taking an exploratory swim around, apparently going nowhere, seeming to check things out – likely an instinctive legacy from the predatory wild, when it was essential to take such precautions. They disappear again, perhaps back into the lodge for a few minutes. Now they're off – first one, then another. A third animal surfaces and heads off purposefully across the broad lake to a predetermined forage site. Eventually we watch four beavers disperse into the lake and we split up to follow them.

Duncan and Hermione go to a place where we have already found a recently felled birch. An animal heads straight for it and meets them there, unconcerned by our scent and that of the dozens of other people who have passed by that day. I stalk my animal, guessing where it will make a landfall and trying to get there first. I have to wade through a marsh. Most of the trees bear the chisel marks of teeth, but few are felled. The feeding here is good all summer: herbs and forbs in abundance on the shore and in the wetland. For seven months of winter the lake freezes and the beavers drop down into semi-hibernation, living off bark from logs stored underwater around the lodge, sleepily harboured in their snug chamber for days at a time, their feeding activity beneath the ice reduced to a survival minimum.

My beaver is not so obliging. It comes ashore and sees me immediately. From a few feet away it eyes me coldly, as if to say, 'Why don't you go and play golf?' I decline, but it chooses not to be watched so closely and slides away through the silvery shallows, diving and disappearing as silently and purposefully as it appeared. I go back to the path to meet up with the others and to share observations. As we return to Songli, we begin to feel that we know a little of the European beaver, understand some of his ways.

Our time at Songli is brief, but it has been action-packed. We have seen three-toed woodpeckers chiselling the royal flagpole; fieldfares and redwings have streamed through in full breeding plumage; high overhead cranes have loosed their haunting cries into the streaming dawn. Red-deer hinds have timidly stepped out of the woods to graze the headlands of stony fields and we have sallied forth in the half-light of dawn to find the other great deer of the Norwegian forests, *Alces* the elk, now better known by its Canadian name, the moose.

For this little expedition Hermione set her alarm and woke first. Blearily she wandered through to the kitchen to get a drink of water. Standing at the sink, looking out over the green sward, she became aware of a movement at the forest edge. The first I knew of it was being shaken awake. 'Daddy! Daddy! Wake up! There's a moose outside!'

And there it was. There is nothing quite like a moose, especially a large bull moose, in full, spreading antler. It is enormous and prehistoric, seeming to belong to another, ancient world of mammoths and cave bears. Like the call of the cranes, there is something of our own wild and indecipherable past in its great shoulders and long, elephantine nose beneath that huge crown of spiky bone. This fellow was magnificent, eliciting instant respect. He was testing the wind before crossing the Songli lawns on his way down to the lake. He stood stock still, the charcoal of his back barely visible against the shadowy conifers behind, his legs fine and pale like young birch stems.

Hermione had fetched her binoculars and was studying him closely. He tiptoed tentatively forward, coming closer so that she caught her breath. He passed the end of the cabin and we nipped through to the living room to see him emerge again only a few yards away. His great bulk seemed to glide effortlessly forward on long, refined legs. 'He's so graceful,' she observed, as he drifted elegantly away from us. Just once he turned and looked back at the house, as if he knew we were there. We could see into his dark pond of an eye; an ear flickered beneath the clawing, upturned fingers of heavy palmed antlers, and the grey folds of his dewlap swayed softly to the movement of his head. The down-arching nose quivered momentarily, and then he turned and melted into the birches, a compelling silhouette of old Europe, a lasting image to seal our time in this wild and beautiful place. 'Wow!' she said softly.

On our last evening we decide to return to the Songli marsh for one final look. We creep past the beaver dam. There is fresh felling going on downstream; we think we might try there. We settle. As the light dwindles the stirring summons of cranes yodels up to us from far away on the lakeshore. High above us, invisible, we hear the continuous, staccato quacking of red-throated divers bickering across the sky. Somewhere in the distance a pair of ravens croak scurrilously as they hurry to their mountain roost. Mosquitoes are troubling Hermione. I have to remind myself that she is only ten. Sometimes it's too much to expect her to apply the gritty discipline of a hardened naturalist. I decide to withdraw with her so that Duncan can keep his own vigil for a better chance of success. We steal away.

When we are a hundred yards clear we follow a moose trail down towards the lake. We can walk easily here. The land is dry and Scots pines have taken over from the willows and alders of the wetland. Thin heather muffles our feet as we approach the marsh. The moon is up and a broad band of quicksilver stripes the brooding darkness of the lake. Herons slice the cool air; sharp and clear, wigeon-whistles rise from the delta ahead of us. Nocturnal nature is out and dancing. The night is opening up like a street market setting out its stalls.

We are clear of the trees now and the moonlight is like pale sherry, the lake a silver tray. We walk hand in hand, quietly, measuring our footfall, unspeaking because there is too much to listen to. Our senses are singing with the stars. With a squeeze of Hermione's fingers I can grab her attention at any time. Ahead of us a dark domed shape looms from the flat marsh. I stop and scan it with my binoculars. I know what it is. It is unmistakable. We walk on. I squeeze and press a finger to my lips. Hermione's eyes are wide-rimmed with expectation; adrenalin shivers through her small hand. It is a beaver lodge.

As we approach it the sedgy scrub is criss-crossed with paths and tunnels; mudslides rip deeply into the stream bank. We sit on a log obligingly felled by the beavers, right at the water's edge. The lodge stands about five feet high from the bank. It is a tangled mass of logs and twiggery, all clogged with mud. Whole branches stick out from its sides like a bonfire; a litter of small sticks surrounds it as if they have been stored there for later repairs. If we sit quietly here for a while we may see some activity. On the edge of the wood somewhere behind us a robin insists on the last word.

Hermione's hand tightens – I have heard it too. Below us, under the bank, only a few feet away, something is moving in the water. It must be a beaver. There is a muted splashing. We crane our necks. Wavelets ripple out into the stream. Whatever is there is big enough to stir the water into a mild commotion. The noise stops. The pool settles. We can't see anything. Suddenly my hand is gripped again. Her keen eyes have seen a bow wave streaming diagonally across the stream, heading straight for us. It has to be a beaver. I can feel the excitement in her taut fingers. It surfaces under the bank, out of our sight. We hear it shake like a dog, then there is a muted sneeze, a snort, as it clears water from its nose. Irresistibly, a smile spreads across my face. I know the author of that short, definitive exhalation. I know it well.

The night air parts like a curtain at the water's edge and the impertinent whiskery face of an otter pops up over the edge of the bank. He is six feet away. His nose shines and twitches. He flows up on to the bank, his hunched back and long tail as sinuous as an eel. Something's not right. He looks uncertain. A floating molecule of our scent must have reached him and he can't quite believe it – people rarely come to this marsh. He rises up on his hind legs. Leaning on to his thick tail as a prop, he stretches himself as tall as he can. His blunt questing snout scans the air, muzzle whiskers – his *vibrissae* – spread like a fan. He is begging for information; his front paws hang like a performing dog. His eyes shine like beads with a glint of moonshine. Whatever intelligence the night air imparts does not please him. Turning on his tail, in a silent twist of wet muscle, he slips back down the bank and into the water. We stand up. The rings of a moonlit target lap the muddy bank and are tugged away by the quiet current. He has gone.

We now influence or control most of the world's wildlife habitat, and all too often our actions fail even to acknowledge other creatures' right to some space on this crowded planet. At best we would seek to put them where we want them – where it pleases us – rather than to allow them freedom of choice. The Norwegian beavers here at Songli have been reintroduced after extinction by over-hunting in the nineteenth century, which has been a great success in terms of conservation and restoration ecology.

The shy and elusive otter has managed to hang on in his secret watery underworld, although pollution of freshwater systems has driven him to the outer edges of his former habitat throughout most countries in Europe. Only in my lifetime, as we have acknowledged our folly and begun to clean up our rivers and streams, has he managed to stage a gradual recovery.

But neither the beaver, graciously permitted, nor the otter, defiantly persistent, is conscious of this human manipulation. Perhaps that is what has pleased us so much. Tonight we have been uninvited guests in their own private, wild world of precious wetland – a wild world where we can never hope to go. We can afford them the space they need, and hope to meet them face-to-face once in a while, but there it ends. In its inimitable, triumphant way nature has drawn a line – quietly said goodnight.

We turn to go home. A figure looms out of the darkness. It is Duncan. His face is alight with triumphs of his own.

THIRTEEN

Svalbard and the Polar Bears

> The truth is that we have never conquered the world, never understood it; we only think we have control. We do not even know why we respond in a certain way to other organisms and need them in diverse ways so deeply.
>
> EDWARD O. WILSON, 1929–

The day before Hermione went off to boarding school for the first time, we went and sat on the beach, her choice on the last day of the holidays – a special day, a special choice – a day for conceding whims. The beach seemed the right place to be, on the east coast of the Moray Firth, close to home, familiar and embracing, where we had been a hundred times before on fossiling and beach-combing forays, dawdling, picking a shell here, a flat stone there: stones I would throw for her, skimming the water and bouncing – never more than twelve times, she insisted, although I am sure I have counted thirteen – to cries of delight and laughter, spontaneous coda to a happy day. When it was time to turn for home, back to the mountains only half an hour away, we were both reluctant to get up and go.

These last moments were precious. They fell quiet and thought-laden so that I could sense their growing burden, the weight of emerging apprehensions unspoken. I tossed my pebbles listlessly now. Common terns flickered elegantly by in little mobs a few yards out to sea, their familiar scratchy cries grazing the surface of our consciousness. Hermione was fingering the hollow carapace of a small spider crab, pink and bleached, dry and brittle; the spiky face and beady eyes on stalks still glared out at us, bitter and defiant. Its body had no nervous ganglia, no bubble-blowing mandibles, no claws or legs or gut, no reproductive matrix, just the shell of a truncated life. Even the barnacles that had hitched a ride on its back were shallow craters of emptiness. She turned it over and over in her hands. Its hollowness seemed to mirror our feelings. We had, for a moment, run out of things to say.

I remembered going off to boarding school for the first time myself, all those decades ago. I felt a disbelief back then that such a thing could ever be good for me; a sense of being abandoned, even banished; of living out the final act of the wonderful drama of freedom and exploration on my own terms in which I had immersed myself to hide from the incomprehensible fact of my mother's terminal illness. Surely, I now struggled to persuade myself, this was very different: we had offered Hermione a choice, the chance to stay at a day school nearer home. But her six adult siblings had all boarded in their time and had pressed home assurances that it offered up many opportunities she might otherwise not enjoy. Borne up by their collective enthusiasm she had bought into the dream from afar, not properly thinking through the more immediate implications of the move. Now, as the last hours of her summer holiday drifted away from her like morning mist lifts from a river revealing unmapped rapids ahead, she had suddenly seen the rocks: the separation from home, the indulgences cut short, the unthinkable distance from her two beloved Jack Russells, Ruff and Tumble, and the end – at least from the perspective of a twelve-year-old – of our forays together into the wild, our triumphant raids on the kaleidoscopic treasury of nature.

For my own part I was wretched. I was losing my beloved companion, my partner in discovery, the girl who had so often drawn me back to the lost world of my own childhood, who had revealed wonder to me all over again, shared my dreams, laughed with me in the eternal sunshine of innocence, never questioning my motives, never doubting my love. I felt like the spider crab, remorselessly hollowed out by the wind and the waves, gutless and bleached. I stared into the waves with the same unseeing stare.

We had achieved so much. My mind skimmed back over the bright encounters of the previous six years: the caterpillars, the adders and slow worms; the toads; the geckos and scorpions; the rhinos and elephants and the roaring lions; the fossils and coral reefs and their sullen octopuses; the jackdaws, starlings and storm petrels; the stags and seals; the beavers and moose and the luminous otter, and many, many more. I smiled; the palpable warmth of those memories began to melt my gloom. Out of nowhere, emerging anonymously from the shining calm of the Moray Firth, a question was forming itself and came bobbing ashore like a little wavelet, catching me entirely unawares. Before I could check myself, the words were out of my mouth: 'If I could show you any wild animal in the whole wide world, what would you most like to see?'

Carefully she placed the crab on a stone and turned to look at me. Her eyes flashed with that old excitement, the precursor to so many happy expeditions. 'Do you mean it?' she asked. As she did so I knew that I had blown it – no going back from here. I thought for a moment. She had once said that she would love to see baby turtles hatch. Up loomed the vision of a sun-kissed tropical island, a sparkling turquoise sea and a friendly local leading us out at dusk, barefoot, hand in hand, along some lonely beach of shining coral sand to witness the prehistoric mystery of giant leatherback turtles lurching ashore to lay their eggs. 'Yes, of course,' I said beneficently.

'Really? Really *anything*?' The disbelief in her voice matched her widening eyes. 'Promise?'

'Yes, of course,' I repeated, adding indignantly, 'When have I ever not meant it?' There was a moment of silence as we stared into each other's eyes, mine seeing only turtles, mischief dancing in hers and a smile teasing her sun-blushed cheeks. 'Polar bears,' she said.

Ice.

Abandon all former notions of frozen water. Forget the opaque cubes in your drink, the crusting on the walls of your fridge, the crisp winter lawn, dismiss even the Christmas icicles glinting prettily from your gutter. Clear your mind of all domestic associations: burst pipes, road gritters, tipping out the frozen birdbath, banish even the immortal image of Pieter Breughel the Younger's ice skaters. All these are tame: ice on the uppermost fringes of cold; water that has temporarily transmuted into something drab, colourless and, more often than not, a nuisance.

Where we found ourselves standing on a frozen arm of the Barents Sea, the nine members of our little expedition, on that unforgettable, shining April day, was as far from domesticity as the imagination can stretch. Ice reigned over this clean, gleaming world as the sand rules the Sahara. In this extreme place ice was the ultimate force, subjugating the rock and the ocean and corrupting the properties of the very air we breathed. It stopped thought in its tracks, demanding attention, dominating all perspective to its own relentless and merciless grip. From the tiniest snow crystals beneath our feet, to the towering crags of glowing icebergs, it freeze-branded its omnipotence into our rapidly emptying brains.

To search for Hermione's polar bears we had flown to Svalbard: a mountainous archipelago in the Arctic Ocean four hundred miles north of the Norwegian mainland and some seven hundred miles north of the Arctic Circle. Of its four main islands, Spitsbergen, at 78° north, is the largest and best known, also containing the islands' only real township, Longyearbyen. There we had arranged to meet two local guides, Jens Abild, a quiet, gentle Dane of 32, who has been intoxicated with the Arctic for twelve years and runs his own dog-sledding business, and his helper, Lisa Ström, a perpetually smiling 26-year-old Swede, who is a student of Arctic technology, researching the impact of PCB pollution in mammals of the polar regions.

During the six months from Hermione's shock announcement on the Scottish beach, our expedition's numbers had swelled. This time Lucy was certainly not going to be left out, and my daughter Amelia (26) who was at that moment wrestling with a restoration ecology PhD at Cambridge and had often accompanied us before on our wild forays, was also very keen to come. Then I asked Paul Ramsay and his wife Louise if they would like to join us with their children, Sophie (20), Adam (17) and George (14) – a Scottish family every bit as philosophically engaged with wildlife and nature conservation as my own.

After a night in a hotel in Longyearbyen – the haphazard, linear assemblage of houses and shops that constitutes the closest town to the North Pole, but actually the size of a modest British village – Jens mounted the nine of us on skidoos and we set off across the crystalline, glaciated snowscape. Our impromptu polar expedition was under way.

Svalbard in winter (April is unequivocally winter, even though the long months of darkness are over) is not for the faint-hearted. Despite the best efforts of the Gulf Stream, which wraps round the top of Norway and keeps the west coast of Svalbard several degrees warmer than the east, the April temperature commonly drops to –35° C, and even in strong sunshine at midday it never rises above freezing. In our seven days here it was to average –18° C. If a breeze whips in from the north or east, wind-chill can rip it down to –40° C in less time than it takes to remove your mitten and pull your zipper-toggle tighter under your chin.

All flesh must be covered. In Longyearbyen you can buy a soft neoprene mask that fits snugly over your nose and lips, with small holes for air. You tuck this inside your woollen balaclava and then your goggles close off your eyes and brow line. Your hat has to have brow and earmuffs. You need layer upon layer of cosy underclothes: long johns, long-sleeved vests, silk roll-necks, thick thermal stockings, under-gloves, then padded trousers and pullovers, and last, but vitally, a voluminous quilted, windproof skidoo suit which zips up the front. When you have hauled on thickly insulated rubber boots and long, fur-lined leather mittens, you are ready to step outside. In driving snow there is a fur-rimmed hood to erect as well. You are locked in, identification no longer possible. Your name has gone, smothered along with your personality and most of your bodily functions, trapped beneath all these crucial defences. We stumbled about trying to familiarise ourselves with this new attire, peering into each other's goggles to see who lurked inside.

We had chosen early April because that is when the female polar bears emerge from their snow dens, bringing their cubs into daylight for the first time at about three months old – perfect little white bears fully equipped for the rigours of the polar elements. Our aim was to find the tracks of such a mother; our dream was that we might be rewarded with a sighting of this special moment in the ice bears' calendar. We laid out our ambitions to Jens and placed ourselves entirely in his hands – there was never any question of wandering about on our own. Not only was this extremely testing terrain, fraught with potentially fatal hazards, like falling through the ice on the fjord where one of the four species of seal found in these waters (ringed, harp, hooded and bearded), has kept it thin at a breathing hole, or breaking through the snow crust on a glacier and disappearing into a deep crevasse, but it is also the most dangerous moment in the year to meet a polar bear. They have little fear of humans at the best of times, but with young cubs the she-bear is guaranteed to be aggressively defensive. You have to be armed. The laws laid down by Svalbard's Norwegian Governor (and haphazardly enforced by his five policemen) demand that you or your guide carry a rifle, flares and thunderflashes for scaring bears. It is also illegal to disturb or kill polar bears unless *in extremis*.

In Svalbard's long, messy and unhappy past (it is first mentioned in 1194 in Icelandic sagas and was occupied by Russian hunters from the sixteenth century), the bears and other fur- or oil-producing animals such as whales, walrus and seals, reindeer and Arctic foxes have been mercilessly hunted and trapped. By the nineteenth century the bear and walrus populations had been depleted to the brink of local extermination in this relatively accessible swathe of the Arctic Basin. Up to 300 bears were shot on Svalbard each year right into the 1960s. Protection began with quotas in 1970 until, in 1976, all the Arctic nations with polar-bear populations signed up to a treaty on total protection, with the limited exception of the indigenous people's (Inuit) hunting rights.

Nowadays the bear is fully protected in Svalbard; the authorities remove sick or injured animals and those few that display an unhealthy interest in human activities. Consequently their numbers have recovered well: the present-day Svalbard/Barents Sea population is estimated to stand at around 5000 (of a world population of perhaps 30,000).

Jens always carried a handarm in a sealskin holster and cylindrical flares in a cartridge belt, fired from a pistol like the old military Verey Light. Lisa travelled with an aged and battered Mauser slung, brigand-style, across her back and I carried six British-Army issue thunderflashes dispersed among the pockets of my suit. I noticed that Lisa had a piece of sticking plaster over the end of the barrel of her rifle to prevent snow clogging it. So simple and effective, it seemed to me to reflect the fundamental practicality of life in the Arctic. That's what the far north does to you: you just do what works – no frills – driven by necessity; imperatives imposed by the conditions and the severely restricted options. Mistakes in this climate can easily be fatal.

But first we had to get there, out to the east coast, on to the pack-ice between Spitsbergen and the smaller island of Barentsøya, where there would be the greatest chance of meeting up with bears – a journey of more than eighty miles, across the roadless frozen wastes.

By the grace of some imaginative Russian, a meteorological research ship called the *Viktor Buynitskiy* had been intentionally frozen into the long sea arm of Templefjord some thirty miles out from Longyearbyen, where it thoughtfully offered its small number of spare cabins and the services of its cheery crew to Arctic adventurers like us. We first saw the ship from afar, from a mountain ridge we had to cross 2000 feet above the fjord, a tiny shadowy blip lost in the white vastness of frozen sea, still five miles away. Not until we arrived alongside her and clambered up the gangway with our rucksacks did we believe that she was real. Her generators throbbed enticingly and her parka-wrapped crew welcomed us aboard like long-lost shipmates.

We spent four nights aboard this friendly ship: warm, comfortable, well-fed and cared for. Each morning we set out on our skidoos to search for bear tracks, staying out until the sun went down. The ship was still 45 miles from the east coast, so getting there was a long and arduous journey. Slowly our elongated convoy of eight skidoos wound through snow-filled valleys, bumped across lumpy glaciers, crunched over gritty moraines and slid past domed pingoes bulging with melt-water ice of sparkling mineral green. Every half hour we stopped so that we could stretch our aching shoulders and legs and sip warm drinks from our flasks. Rime encrusted our faces and our clothes. Our breath froze to our balaclavas, rendering the wool stiff and rough. Icicles gripped our eyebrows. We were warm inside our suits, but our extremities quickly chilled. Lisa rubbed feeling back into Hermione's hands and feet and Jens gave her chemical hand-warmers in little sachets to place under her toes and inside her gloves. We kept our energy up with crunchy chunks of Norwegian fruit-and-nut Firkløver chocolate, until our little convoy set off again – there was no question of standing still for long.

Lucy and I had worried about the effects of the cold on Hermione. Extremes cannot be explained in advance, least of all to a child. We knew we would just have to wait and see. In the event, although she did get very cold fingers from time to time, and even a touch of frostbite on the end of her nose, of all nine of us she was the most at home in the snow, the least curtailed by her cumbersome protective clothing, and the most adaptable to the hardships of Arctic life. We should have known better. As with the searing heat of the Kalahari Desert, she adjusted, made no fuss and missed nothing.

On that third day, 1 April 2003, we found tracks – unmistakable tracks. This was no prank; a gargantuan had passed this way, fooling nobody. We stopped and stared. Awe settled around us in drifts. Spreading my fingers as wide as my mitten would allow, I placed my whole hand entirely inside the circle of one great furry foot. It was thirteen inches long. Hermione could not stretch her pace to match one front-to-rear step of this great bear. I stood and watched her trying, jumping from print to print, and then rolling her shoulders as if to become the bear itself. 'Male,' said Jens without emotion – the stoical, economical way of his race. 'Very big.' The Arctic never wastes words, but we could sense that even he was impressed. We glanced nervously around us. Still smiling beneath her rime-fringed mask, Lisa was eyeing up the landscape. 'Stay with me,' Jens added sternly to us all. 'And do not move far from your skidoo.'

Earlier, in preparation for this moment, he had insisted on a safety procedure that could save lives. Whenever we stopped we were to place the skidoos close together in a line facing the best escape route to open country, his machine first, Lisa's at the rear. Ignition switches were to be left on and the stop buttons raised for a quick, certain start.

We were approaching the sea, coming down off a glacier, but still in a high and exposed position. The wind had swept the ice clear of snow, leaving it clinging to ridges in triangular drifts and skiffs, as sand on a beach collects behind a stone. The huge prints had begun to fill. We walked slowly alongside, our heavy footfalls emulating the leisurely sway of the bear. It didn't seem right to tread in his solemn signature – we felt an authority we dared not defile. 'How long?' I asked. Jens had lifted his goggles so that I could read his grey-blue Scandinavian eyes. He was lighting his curved Sherlock Holmes pipe, balaclava tugged down at the corner of his mouth: a curl of aromatic blue smoke lifted lazily into the freezing air. He shrugged his shoulders, staring at the prints. 'Half a day,' he said. Then, after a pause he added, 'Maybe not so much – a few hours.'

We stood in silence, still looking around. At the glacier's edge the scoured rock of the valley wall showed through the thin snow as a patchwork of dark intrusions. They broke up the gleam of the sunlit slopes in the way that patches of dark fur infuse a white dog's coat. Huge rocks had toppled from the steep slopes as the glacier slowly relaxed its grip, melting its way back into the mountains. They lay about at the ice edge like ramparts ineptly built, or a chaos of ruined buildings left by a departing army, strangely incongruous and imposed. Snow had half filled the gaps, welding them together into long, ragged lines, crudely completing the job and rounding the hollows so that deep shadows lurked there, shadows both deceptively cavernous and inherently sinister – their very innocence tainted with foreboding so that our eyes returned to them over and over again.

Hermione took my hand; even through thick mittens I could feel the spiralling apprehension within us both. Lucy and Amelia came across and joined us. They didn't need to say anything. Following Jens' lead we stood and scoured these luminous darknesses through binoculars until the cold slicing into the flesh exposed around our eyes became intolerable. Nothing moved. After the roar of the skidoo engines the land was eerily silent, but it was a silence so loaded with prescience that it seemed to sing. Our bear had, it seemed, moved on. It had imposed its fearsome gravity upon the snow in a slow, rolling gait, a graphic so eloquent that its essence seemed to have remained, like the awful visitation of a displeased god, invisible and yet everywhere about us, a lingering presence that touched us all. I knew I would never forget those tracks.

A high-pressure zone had parked itself over the islands. The sky was clear and blue. The sun shone brightly from an elevation 20° above eye level, strong and unquestionably on our side, but still ineffectual against the searing cold. Even in full sun at midday the temperature remained stolidly at –17° C. There was no wind; a small, momentary mercy from whatever deity ruled this lonely place. To the south, a distant striation of cirrus cloud greying the horizon suggested that change could be approaching, but it was far off and seemed entirely benign. Imaginations still spiralling, we sped on towards the frozen ocean with surging spirits.

For the second time that morning Jens' left hand signalled a halt. We formed our obligatory rank and checked our safety routine. He was standing beside more tracks that wove a straggling line across our path. He was beaming broadly. Here were the familiar oval indentations of an adult bear, slightly smaller this time, and on either side, irregular-paced, big-dog-sized pugs meandered along, sometimes crossing, then parting again, never straying more than a few feet from the adult trail. There could be no doubt. These marks read like a thriller to our widening eyes, inscribing new excitement into our smiling anticipation: here was our whole purpose for being in this extraordinary polar place – a mother and two cubs. A fizz of electricity sparked through us. Two bear cubs had been skipping and weaving their new lives into this virgin wilderness, every ice rock an excitement, every riffle and snowdrift a polar meadow in which to prance and roll.

'Fresh,' said Jens. 'One hour.' We moved off again, keeping to one side of the trail, all eyes searching, hearts thumping, an adrenalin-aroused wariness tingling in our veins. Soon we crossed from land on to the sea. There was no obvious transition, no shoreline where you could definitively say the land ended and the ocean began, just a gradual flattening of the landscape to a broad, bumpy, ice-strewn expanse stretching away from us to the east. The horizon was formed by the dull blue hump of the island of Barentsøya. In the middle distance towering icebergs studded the surface of the frozen sea, while to our left stretched the luminous face of the glacier, its sheer end bright aquamarine where each summer it met open saltwater and the opposing forces of gravity and buoyancy clashed, fracturing its mass and calving its offspring into the ocean.

Only minutes later up went the arm again. We pulled alongside and dismounted. The smile was broader now. Out came the pipe and again the friendly fragrance of tobacco caressed our chilled noses. 'This is good,' he said, walking forward to a mound in the ice. We stood looking into a hole. The snow around was a broken confusion of prints, large and small. The adult bear had been digging here. 'She smells seal,' Jens told us, following the prints towards a further mound. More digging, more chaotic trampling of snow and then, with all the suddenness of an unexpected gunshot, we saw daubs of frozen blood smeared and splattered around the interior of the hole. The excavated crater was four feet wide and three deep. Jens produced a hand shovel from his skidoo and dug beside the bloodstains. In a few seconds he had struck open water. 'It is very fresh. No time to freeze again,' he revealed.

This pore in the sea's rigid skin was a ringed seal's breathing hole. Beside it there was a hollowed-out den in the covering snow-drift where she had given birth. The seal pup had been hidden there in a tiny declivity the size of a dog basket, tucked beneath the snow surface. There, unfearing, unknowing and utterly defenceless, it slept out the long hours between feeds of fat-rich milk from its mother, who returned only infrequently in case her scent gave the location away.

Nature has equipped the ringed seal pup with two principal mechanisms to cope with polar-bear predation: huge numbers and these shallow hideaways. Ringed seals are among the most numerous seal species in the world. There are perhaps some three to six million individuals spread around the northern circumpolar seas. Consequently, the percentage loss to predation is relatively small, although there can be no doubt that bear pressure has been a major factor in shaping their biology and their behaviour. Without these snow dens to hide them, the bears would simply travel from one exposed pup to the next, killing them all. Of the two defences, the first is the only reliable one for the species because in its opportunistic, perpetually testing and probing way, nature has equipped their arch predator with a nose so keen that it can smell out the presence of a luckless individual seal beneath the ice many yards away. Once located, the game is up for the seal pup.

That first excavation we found was an exploratory dig by the hungry mother bear; she had then adjusted her bearings and homed in on the helpless pup in its shelter a few yards away, scouring away the snow and ice with her shovel-like forepads and long claws. In seconds the pup was gone – death at one crunch from mechanically powerful jaws, torn to hot, gulpable chunks in a cloud of ursine breath. Everything had been consumed. There was no shred of fur, no bone, no claw or skull, no smear of congealed blubber, just the sombre inscription of a shattered life on the crumpled snow quilt, nutrients recycled even before they could cool. The single-file tracks of an Arctic fox wove a cautious exploration around the scene. This time he went hungry, there was nothing left to scavenge.

I began to feel that we were going to find our bears, to realise our dream. In such excellent conditions, with visibility crystal clear for several miles, the chances were running high. A strange feeling of calm came over me, despite the excitement of the moment. It was a sense of anticipated fulfilment, that we were well on the way to achieving our goal. Not just because we had come to this extraordinary place and coped with the cold and the other discomforts, nor because of the success of the expedition so far, but because of a sensation that spun me back to that day on the Moray Firth beach, the day when Hermione first came up with the whole idea.

I had wanted to do something special with her to close this particular chapter of our lives. It wasn't that I thought we wouldn't be doing exciting things together in the future, or that by going off to a new school Hermione would somehow abandon her childhood, but I knew that priorities would inevitably shift, and I wanted to allow that to happen against some vivid memory of her own devising. That, after all, was why we were here. The polar bears were her choice, although I had always harboured a similar dream myself. Now we were on the verge of catching up with this she-bear and her cubs. We had found her tracks and her kill. The chapter was, it seemed, coming to a close.

Our little convoy set out again, heading east across the pack ice, weaving slowly and carefully between towering icebergs jumbled together, locked into the surface like Christmas cake decorations. Each of these we had to circumnavigate slowly and well out from their craggy borders. Our she-bear could easily be lying up in the snowdrifts packing the interstices of one of these, resting her cubs after a feed, ever watchful, ever ready to defend her youngsters against the ravening male bears that roamed the pack ice in the perpetual quest for seals. They would not hesitate to kill young bear cubs – even their own – and often do.

After an hour on the sea we stopped for a breather. Twice we circled a colossal berg complex before satisfying Jens that it was safe. We shut down the skidoos; the silence swamped us, descending like a mood. We raised our goggles and pulled up our stiff balaclavas, re-entering the world as if coming out of a stuffy cinema into sunlight. While Jens and Lisa prepared hot drinks we gazed around, mesmerised by this extraordinary, crystalline georama. With Lucy and my two daughters at my side I walked a few yards out into the ice field so that we could look back at our polar caravanserai from a little distance.

As has happened to me many times before in some of the world's wild places, I was gripped by the overpowering contradictions of the scene. Here we were, in a wilderness closer to the North Pole than to any town or city, a wilderness of exquisite natural beauty, immaculate and pristine, into which we had roared on noisy, exhaust-belching machines, dressed from head to foot in artificial fabrics (except for our blissful silk and woollen undergarments), enthusiastically suffering often severe discomfort so that we could experience an extreme variety of nature as she really is. It is a personal paradox I have never quite managed to reconcile within myself – not so much a guilt as a curious sensation of voyeurism, which imposes upon me a melancholic acceptance that modern man *is* now estranged from nature and that all we can do is stand and wonder. I wanted to explain this to the girls, but it seemed incongruous so I held my peace, only muttering something fatuous like, 'Try to remember these moments'.

There are some things in life one never forgets: images seared into the brain and sealed there like a brand. That iceberg and its encompassing vista was one of these. Its towering mass was no colour I had ever witnessed before, a rare Chartreuse green with a splash of milk, beryl deepening to jade in the shadows and a shining turquoise enamel where the sun collided and refracted across its clear, cut faces. Its hard geometric profile seemed to laze against a soft horizon of pink and blue snow-quilted hills, like a cluster of rare jewels on a velvet cushion. To our left a long glacier cliff of nacreous turquoise glass, from which our bergs had calved, shone like the walls of some fantastic fairytale citadel. Out to sea many more icebergs littered the view, their unpredictable size fooling with perspective, so that we were unable to judge distances at all. And there, ranked like leathered bikies on a seaside promenade, our ski-doos sat, garish in black and red and royal blue, with our black-suited colleagues, standing out as hard and intrusive as ink spilled on a page.

Rested and refreshed we moved off again in single file. Jens had impressed upon us to stay close; we all knew that our she-bear was out there somewhere. A mile passed, then another, as we weaved our slow mechanical trail round iceberg after iceberg, always keeping a safe distance out into the pack ice, casting round, searching, searching . . . Suddenly Jens' arm was raised and he was waving and pointing. We pulled up beside him and followed his line.

She wasn't white, she was yellow, the yellow of old candle wax. There she stood. There was our she-bear, and beside her the two small cubs whose puppy prints we had followed into this glistening wonderland. We pulled out our binoculars, kept warm inside our suits to prevent condensation in the cold air. She was two hundred yards away – a safe distance – and retreating slowly. She had scented us and then seen us (she could hardly miss us), so she had decided to take her cubs away from the man-smell and the noisy machines. She was heading not away, but obliquely to the left of our vision, across the open pack ice, stopping every few paces to raise her great head high in the air, her nose sampling the wind like a gundog. I moved closer to Hermione, who was intent, locked within the rings of her binoculars. 'Is that what you wanted?' I asked quietly.

'Look at her raising her head . . .' she replied, still staring out into the whiteness, '. . . and those cubs . . .'

Many more words would come later, back at camp, at the airport, at home, months, years later when we could relive these moments and the timeless, electric memories they would forever spark.

That is the image – that's the one I have locked in for good – this great polar predator raising her noble head high above her back while her cubs trotted at her side, and Hermione quietly watching. They were slightly behind, one on either side of her, meandering along, contented, full of milk, secure, unconsciously learning; the great white hunters no longer hunted, but as they were meant to be, exquisitely wild and unconditionally free.

Envoi

We are what the atmosphere is, transparent, receptive,
pervious, impervious,
We are snow, rain, cold, darkness, we are each a product and
influence of the globe,
We have circled and circled till we have arrived home again,
we two,
We have voided all but freedom and all but our own joy.

WALT WHITMAN, 1819–92

Suddenly Hermione was twelve. The previous September we had taken her off to boarding school for the first time. We drove her there, of course, Lucy and I, down the long snaking highway south, high over the Slocht Pass through the Monadhliaths and sweeping across the broad-shouldered Drumochter, through wide, whispering Grampian moors where the last of the heather lent a lilac haze to the late summer afternoon; a drive of two and a half hours, away from the softer, friendlier hills and wooded glens of home. We were unusually subdued. When we arrived we felt flat and powerless, victims of our own actions.

We helped her unpack her things, hung up her clothes in the wardrobe in an artificial orderliness we knew would not last. We placed her family photos on the locker top and made polite but distinctly stilted conversation with the other three new girls who were to share her functional study-bedroom. The time came to go. We had been dreading it for days. We stood about forcing smiles and saying inane things such as, 'What a nice view you've got from your window.' Suddenly Hermione turned to me and said quietly, 'Daddy, why don't you both just go.'

It was an even more sombre journey home. We felt like traitors – at least that's what we convinced ourselves – but the truth probably is that we were far sorrier for ourselves than for her; simply bereft of our last shot at being twenty-four-hour, hands-on parents. Even though we had done it before, it was no easier this last time. In our different ways it was a huge wrench for us both. For Lucy it marked the close of that wonderfully special phase of dependent motherhood. For me, it felt like the last summer holidays of our special companionship, not just on a weekend or holiday basis, but day in and day out after school – Hermione as a luminous, galvanising presence always there, always smiling and interested, always keen to come if I was nipping off to check a recently discovered osprey's nest or to investigate a report of a wildcat killed on the road further up the glen. It was the end of long summer evenings together in this enchanted land when, through May and June and into July the daylight dawdles on and on, handing us a whole added day after school was over; and the end of those sharp, white winter mornings when on our way to school we would venture out in the glowing dawn to see if the pine marten had been round in the night, his fresh footprints sprinkled across snow-quilted lawns.

Now, all these would somehow have to be crammed into rushed exeats and shoehorned into holidays crowded with all the inevitable distractions of growing up. For me it was a time of unspoken trepidation, culminating in that final precious day on the Moray Firth beach, the day of the spider-crab carapace and the terns flickering past us over the gentle waves – the day of the polar bear.

Although at the time it had dumbfounded me, the polar bear had been a masterstroke. Lifting us out of our gloom in an instant, we had begun to plan. There was no chance of finding bears before winter, so we had six months in which to research our next great expedition, at Easter, perhaps sometime in March or April. Would it be Alaska or Greenland, or Churchill on Hudson Bay, where polar bears congregate, attracted by the rubbish in the town's dump? Churchill is the inverted, zoo-like ecotourism Mecca for polar bears, where tourists are securely locked inside armoured vehicles that grind through the rubbish to bring their camera-snapping occupants face to face with the bears. Or could we conceivably get out to Ellesmere Island, or even Franz Joseph Land? Was there a NATO or military base somewhere on the pack ice that we could persuade to help us, in whose warmth and sanctuary we could spread our bedrolls? And when exactly would the female bears emerge from their dens? Would the denning and cub-emergence dates be the same for Prudhoe Bay on the Beaufort Sea as for the scattered islands of the Barents Plain?

Just when would the grip of winter relax sufficiently to allow a twelve-year-old and her father to get to the Arctic and somehow survive long enough to witness this climactic moment of extreme wildness we had now both set our hearts upon? How long would we need to stay there to have a fair chance of success? How could we ensure our safety? Who were the experts? Where would we find a local guide? Who else in the family might be mad enough go with us?

Right through that winter we plotted and planned. Letters, maps, photocopied extracts from learned Arctic books and complicated lists of equipment flew back and forward between school and home. Whole schemes were taken to the brink of fruition and then ditched. No, the Churchill dump was definitely *not* where we wanted to experience our polar bears; the more we learned about it the less we liked. We sought true Arctic wilderness unsullied by man's ugly footprint. Wherever we ended up, we wanted to afford our bears the dignity their majestic predatory status deserved; for them to be there because they belonged there, not artificially enticed by ugly heaps of smouldering rubbish. And for the same reasons we both agreed that we were ready to suffer a little to achieve our goal, so that we could understand properly how exquisitely adapted the polar bear is, and, if we were lucky, meet his wildness on his terms. Slowly, Svalbard and the Barents Sea began to emerge as the most likely, and, as it turned out, the most accessible place to go. Even the Gulf Stream seemed to be on our side.

In the end my fears were unfounded: I hadn't lost my companion at all. Every day Hermione seemed to be there, with a letter or a phone call or a text message; always planning, asking, probing: 'Come on, Dad. Hurry up and get your act together!' As the weeks passed I forgot my doubts. A private code emerged between us: all her letters and texts ended with the same cryptic message – 'MORE!' It seemed to sum us up, encapsulating everything she had to say, everything I wanted to hear. Once again our future opened before us. Those early years of formative childhood had gone, we had moved on. Out of the Arctic expedition and its excitement a new life had emerged and with it a maturing confidence for us both; a growing appreciation that those precious years had been a foundation for life; that they were our shared childhood, to be visited and revisited over and over again, there for good.

As we flew back into Aberdeen airport at the end of our polar expedition, I experienced a curious sense of time not running out, but running in again against my will. The airport clock clicked its endless digital progression, forcing seconds and minutes upon us in a way that had been entirely absent in the Arctic. During one of those wearisome queues that define modern airports I found myself resenting the intrusion of the clock – any clock – as an unjustifiable pretension of our age; resenting the suggestion that a clock can somehow quantify the value of time and experience. At that moment I saw clearly the futility of what, with extraordinary complacency, we have come to accept as normal life: the ordering of our lives with time-driven routines which swamp our deeper consciousness with pointless and meaningless urgency. On its own, punctuality is an empty virtue.

Life is an album; a collection of fragments of time charged with deeply personal sensation and meaning, some lasting seconds, some years, but all defying the orderly definition of time. The greater the reality of the experience, the deeper its impact and the more precious the moment, which the memory revisits over and over again, in laughter or in tears or just in quiet contemplation – stretching time way beyond the pedantic limitations of the clock. In that long week of Arctic undertaking we had watched polar bears for only a few minutes, but the recollection of those images has already filled weeks, months, years – they are locked in for life.

What is love if not time given in joy and delight? Such has been my experience these past six or seven years that sitting down to write about them was irresistible – an opening of the album, flicking through images to relive life's essences. But the sensations these words evoke, the nature's child within us both, are not exclusive; they aren't the private preserve of Hermione and me, nor even the happy accident of timing that allowed us for this handful of all-too-fleeting years to indulge our shared passion for wild things. Rather, they are a joy latent within any child and any parent; they are the birthright of us all, a smiling, backward glance at our primeval origins, at our universal and ancestral history in nature, from whence we all emerged and which shaped us all the way we are.